ASHEVILLE FOOD

ASHEVILLE FOOD

A History of High Country Cuisine

RICK McDANIEL

PHOTOGRAPHY BY AMY KALYN SIMS | FOREWORD BY JULIE STEHLING

AMERICAN PALATE

Published by American Palate
A Division of The History Press
Charleston, SC 29403
www.historypress.net

Copyright © 2013 by Rick McDaniel
All rights reserved

First published 2013

Manufactured in the United States

ISBN 978.1.60949.865.8

Library of Congress CIP data applied for.

Notice: The information in this book is true and complete to the best of our knowledge. It is offered without guarantee on the part of the author or The History Press. The author and The History Press disclaim all liability in connection with the use of this book.

All rights reserved. No part of this book may be reproduced or transmitted in any form whatsoever without prior written permission from the publisher except in the case of brief quotations embodied in critical articles and reviews.

To the chefs, farmers, locals and tourists who make Asheville one of the best places on Earth to live and eat.

CONTENTS

FOREWORD

My husband, John, and I opened Early Girl Eatery in downtown Asheville on October 13, 2001. On October 14, Rick McDaniel walked in our door and ordered his first of many pumpkin gingerbread breakfasts.

Asheville was quieter back then, happening but not bustling. You could count the farmers' markets on one hand and easily know every street musician by name. For John and me, life was very busy, and we were never not cooking, bussing tables or serving guests. Rick was writing for the *Asheville Citizen-Times*.

Rick knew everyone in the restaurant business, and he knew food. He wrote an article on Early Girl's first anniversary that is still the best piece out there to truly capture who we are and what Early Girl set out to be. Rick is passionate about regional food just as you expect a Southerner to be, but more so. He knows his chow chow and his gravy and the history behind every variety out there. His cookbook, *An Irresistible History of Southern Food*, and websites (ChefRick.com and Hushpuppynation.com) are testaments to both his vast knowledge and his irresistible sense of humor. Over the years he has been an active participant and a witness to the blossoming of Asheville's food scene.

During an economic downturn that has left so many towns scrambling, Asheville has defined itself and grown. The community's desire for sustainable food systems and protected landscapes has helped launch nonprofits that help small farms thrive. The collision of local demand and farmers' bounty with the creative culture of independent business has created something worth emulating. Organic growers, hot sauce makers, servers, butchers and food photographers are all making a living working with local resources.

At Early Girl, we have been buying from many of the same farmers for almost twelve years. Working together we now have successful business stories and growing families. In the meantime, more and more chefs have arrived and more farms have grown and prospered. Farmers inspire us, and hopefully we inspire them. Asheville's restaurants deservedly draw rave reviews from national critics. Their success has, in part, been a product of an agricultural infrastructure needed for a seamless farm-to-table culture. The creative collaboration between chefs and growers resulted in a ballooning farm-to-table synergy in Asheville—over the past decade, in particular.

It now takes a full typed page to list all of our region's tailgate markets. John and I now get enough time off to work on opening a new restaurant, King Daddy's Chicken and Waffle. Street musicians I knew by name have put out albums and now populate our dining room next to tourists and council members, all expecting ramps in spring, strawberries in early summer and local, grass-fed beef all year round.

And my friend Rick gets to share the story of how a small Appalachian town came to offer so much exciting food and opportunity. And that's delicious in and of itself.

—JULIE STEHLING
Co-owner, Early Girl Eatery and King Daddy's Chicken and Waffle
Board of Directors, Appalachian Sustainable Agriculture Project

ACKNOWLEDGEMENTS

This book is the result of a carefully planned happy accident.

I had actually started the research for a completely different book when I got a call from my editor at the time, who told me about a series of books The History Press was planning about the local food scene in a series of cities across the country and asked if I would write the one for Asheville.

I have to admit I was a bit reluctant at first. While I thought it was a pretty cool project, I already had my mind wrapped around the other book. The more I thought about how Asheville had changed from a "meat and three (canned) vegetable" town into a true destination for culinary tourism, I realized that almost all of the people who had made that happen were friends of mine, and I was the one who could best tell their story because I already knew the story.

I owe many thanks to Banks Smither, my current editor at The History Press, and to his predecessor Jessica Berzon, who has since moved on to other pastures.

The best thing Banks ever did for me was to hire Amy Kalyn Sims as my photographic collaborator on this project. She is a consummate professional and a completely amazing master of light, shadow and color, and the photographs in this book are amazing.

My wife, Polly, did her usual yeoman service as preliminary editor, head cheerleader and chief talk-Rick-down-from-the-ledge person. She was always there with soothing words of encouragement (and a double Scotch, at times) when I was pacing the floor moaning, "There's no way in the name of God I'm ever gonna finish this thing in time!"

ACKNOWLEDGEMENTS

The story of food in Asheville cannot be told without acknowledging the contributions over more than a century by George Washington Vanderbilt and his Biltmore Estate, the magnificent 250-room mansion he built at the close of the nineteenth century. From state-of-the-art kitchen appliances to elegant Edwardian feasts to pioneering work in sustainable agriculture that is still in practice today, Mr. Vanderbilt, his descendants and the generations of staff at the estate have made their mark on dining in Asheville. LeeAnn Donnelly and Marissa Jamison in the media relations department were great helps in putting me in touch with experts and securing photographs. Darren Poupore, chief curator at Biltmore Estate, and his staff were extremely generous with time and information, as was Estate Historian Bill Alexander. The information in chapter two, "Dining at Biltmore," was almost entirely based on research done by estate curators.

Jessica Furst of the Special Collections staff at the D.H. Ramsey Library at the University of North Carolina–Asheville was extremely helpful in accessing historic photographs from the E.M. Ball Photographic Collection. The collection also held a wonderful oral history from Billie Walker Tingle, whose family were restaurateurs at the dawn of the twentieth century and owned the iconic Tingle's Cafe. This was a treasure-trove of information on early Asheville restaurants.

When I was asked to have someone write a foreword to this book, there were many people who could have done a good job, but there was no one who I felt would understand what made Asheville the unique foodie paradise it has become better than Julie Stehling, co-founder with chef and husband, John, of Early Girl Eatery, Asheville's acclaimed pioneering farm-to-table Southern restaurant.

In addition to being committed to local sourcing even before the doors opened for the first time, Julie has served on the board of the Appalachian Sustainable Agriculture Project and been involved in the Asheville eating renaissance that has transformed us into the dining nirvana we are today.

I would also like to thank the chefs who pioneered Asheville's dining transformation—Mark Rosenstein, Vijay Shastri, Joe Scully, Damien Cavicchi, Laurey Masterton, Brian Sonoskus and others—whose talents and long hours have made an amazing transformation happen.

The new generation of farmers who appear in this book are the driving force behind some of the most creative and dynamic chefs anywhere in the United States. Their commitment to sustainable agriculture and scientific farming has caused a quantum leap in the amounts of fresh, flavorful heirloom fruits and vegetables available at tailgate markets, local grocery stores and the farmers' market and on the tables of Asheville's restaurants. Without them, Asheville would not be the dining destination it has become today.

The art and science of turning a collection of Word documents into a real book can be an aggravating and frustrating experience, but it was a breeze this time thanks in large part to my project editor, Darcy Mahan.

Thanks also to Amber Hacker, Heidi Flick, Nancy Carter Crump, Kyle Brown, Tony Keller, Jeff Jones, Sandie Rhodes and Hilary Dirlam for eagle-eyed proofreading and immoral support. A special thanks also to Céline Lurey of Asheville's awesome Céline and Company Catering, who has been a dear friend and the president of my fan club for many years.

Lastly, an apology. There are a lot of chefs, farmers, artisan bakers and others who deserve to be in this book but for reasons of space, logistics and, sometimes, ignorance on my part of the role they played are not. The best way to find them is through the Local Food Guide published by the Appalachian Sustainable Agriculture Project, which is available locally in printed form at many restaurants or online at asapconnections.org.

INTRODUCTION

This is the somewhat unlikely tale of how a small city nestled in the Blue Ridge Mountains of North Carolina became a food lover's destination, an undiscovered gem that became the darling of the Food Network and the Travel Channel.

Asheville turned into a food Mecca about a decade ago when it became a pioneer for the national farm-to-table movement, and wow, is the secret ever out now. "Cooks from DC to NYC always look at me in amazement and wonder when I tell them I'm from Asheville," said Mike Moore, of Seven Sows Bourbon & Larder. "That's amazing. We have something very unique here, and we should be proud of that."

For those of us who knew and loved Asheville in the 1970s, the idea that the city would be nationally recognized as anything other than a pretty mountain hamlet and the home of Biltmore Estate, George Vanderbilt's 250-room cottage, was pretty far-fetched.

Not that the rest of the world had totally ignored us, mind you. Mr. Vanderbilt managed to keep a steady stream of nineteenth- and early twentieth-century glitterati coming through the tiny railway station in Biltmore Village, where one of the estate's luxurious carriages took the visitors up the long, winding driveway for the beginning of what could be a two- or three-month stay. Once there, guests dined on sumptuous feasts of oysters; wild game from the estate; seafood brought up from Charleston, South Carolina, in refrigerated railway cars; and grain-fed beef from the estate's farmlands.

By the 1920s, tourists from all walks of life had found the secret of Asheville's beautiful scenery and Southern hospitality and were taking their holidays at the Grove Park Inn and other hotels.

By the end of that decade, however, Asheville, like the rest of the country, had fallen on hard times. While the Grove Park Inn survived, and even grew in popularity for the wealthy who could still take vacations in the midst of the Great Depression, most of the grand hotels and their fine restaurants fell victim to the times, leaving Asheville in the culinary doldrums for several decades to come.

The end of the Second World War brought GI Bill money and an influx of fresh blood to the area, and restaurants such as Buck's, Babe Malloy's and Wink's brought American cuisine to its "meat and three" zenith. Even though you could dine on steak and lobster tail at the Sky Club or the Grove Park Inn, to the majority of the world, Asheville was still a culinary secret.

Somewhere along the line, everything changed. As I write this in the spring of 2013, two of Asheville's chefs have been nominated for James Beard awards, the chef's equivalent of the Nobel Prize, and one of their restaurants has been named among the twelve best of 2013 by *GQ Magazine*. Last year, a graduate of our community college's culinary program was named Best Young Chef in the World by the American Culinary Federation. TripAdvisor listed Asheville as a Traveler's Choice Food and Wine Destination, along with New York City, San Francisco, New Orleans and Chicago. The darlings of the Food Network have graced us with their presence, from Bobby Flay to Rachel Ray and several more to boot. And when President Obama makes one of his increasingly frequent visits to Asheville, a large black SUV filled with guys in suits, sunglasses and earpieces heads out for 12 Bones Barbecue in Asheville's River Arts District with a large takeout order for ribs scribbled on Air Force One stationery.

This is not supposed to happen in a town this size.

To have a great restaurant, the first thing you need is, of course, an innovative and passionate chef. But the chef, no matter how good, needs six essential ingredients in order to succeed:

- A good location, preferably with low purchase price or cheap rent
- A good source of labor
- Someone who knows the business end
- A ready supply of fresh ingredients, hopefully locally sourced
- A steady flow of customers
- A way to get the word out

A series of lucky breaks came together in the late 1990s that gave chefs all of these tools and made Asheville's restaurant scene explode.

The city of Asheville had invested heavily in the boom time's stock market during the 1920s, and when the market crashed in 1929, the city went bust. But instead of declaring bankruptcy, Asheville vowed to pay back all the money.

This took decades, and when Washington came knocking in the 1960s and 1970s with matching funds for urban renewal, Asheville took a pass. Thus, all of our beautiful old downtown buildings were spared the wrecking ball that turned Charlotte into architectural vanilla. By the time the restaurant renaissance was ready, there was a huge inventory of downtown storefronts available at bargain basement prices.

Meanwhile, something pretty radical was happening over the hill at the local vocational school, Asheville-Buncombe Technical College (now Asheville-Buncombe Technical Community College, mercifully shortened to A-B Tech from here on out).

A Frenchman named Bob Werth, a classically trained chef, was hard at work trying to convince the A-B Tech board that it needed to start a culinary arts program alongside its plumbing and welding classes. Bob was very convincing, and in 1968, A-B Tech started training chefs. As the newly trained chefs found work all over the country, word spread about the excellent value and strict standards at A-B Tech, and the program soon had students from all over the United States. This gave Asheville restaurants a ready pool of trained kitchen staff.

For a restaurant to succeed, much less flourish, someone in the organization must know and tend to the business end of the business. Marketing, business plan, finances, dealing with bankers—all the stuff chefs don't like to fool with and usually don't know anything about.

In another lucky break, in 1989 a nonprofit called Mountain Microenterprise Fund (now Mountain Bizworks) came along and started classes for entrepreneurs to learn how to start and run a small business. Dozens of chefs and their business partners took advantage of the knowledge, and soon restaurants like Salsa's, Early Girl Eatery, West End Bakery, Zambra, Sunny Point Cafe and Laughing Seed sprang up with a much better chance of survival due to the knowledge and support gained from their association with Mountain Bizworks.

Along with the great chef, good location, kitchen help and business knowledge, the restaurant needs customers, and a steady supply of them. That's where the Asheville Convention and Visitors Bureau came in.

Prior to about 1998, we had a solemn ritual that took place every November. All the citizens of Asheville would gather at Pack Square, and as the last leaf fell, we would wave goodbye to the last tourist as the city crews rolled up the sidewalks. Lots of businesses and restaurants would go to "winter hours," and some would close for a month or more at a time.

Luckily, the Asheville Convention and Visitors Bureau knew this was no way to run a railroad (or a successful tourist destination) and, along with Biltmore and Grove Park, started promotions and activities to make Asheville a four-season destination. Now, we have between three and four million visitors a year, and that number is expected to double in the next couple of years.

As all these disparate puzzle pieces came together in the late 1990s and early 2000s, national media began to notice the unlikely transformation in this sleepy Southern town. *Southern Living*, *GQ*, *Bon Appétit*, *Conde Nast Traveler*, the *New York Times*, CNN—soon everyone had a blurb about the cool restaurants with the hot cuisine springing up in Asheville.

The culinary renaissance was also driven by a population increasingly peopled by folks who moved to Asheville from all over the United States, after we began appearing on everyone's list of the best places to retire and to live. Then a new generation of young chefs and a new generation of college-educated young farmers came together in a culinary synergy that sparked fresh, creative cuisine. Soon the rest of the country started to notice that we had something very cool going on here.

You will hear the term "farm to table" used often in this book. This isn't a new concept—the great chefs of the early twentieth century knew of the special relationship that exists between a chef and his or her garden. But the practice had fallen out of use as more and more chefs in the post–World War II era found their ingredients on the back of food service trucks.

But beginning in the late 1970s and continuing to this day, the farm-to-table culture transformed the way Americans eat. And much of that movement began right here. "Asheville is the Fertile Crescent of farm to table," says Jason Roy, one of the new generation of young chefs you'll meet in this book.

The result of this symbiosis between people, farmers and chefs was an explosion of independent restaurants working with small farmers to make Asheville one of the most unique culinary destinations in the country. This book chronicles how that came to be.

Chapter 1
BEGINNINGS

To understand how Asheville became a food lover's paradise, you have to first look to the mountains. Their story goes back a few million years.

Long before there was a Pack Square, or any restaurants, or even any dinosaurs, there were mountains—tall, majestic mountains, as high as the Rockies or the Alps.

These were the Appalachians, heaved skyward by the greatest tectonic collisions the Earth has ever known.

Running from the Maritime Provinces of Canada all the way to central Alabama like a rocky backbone, the Appalachians consist of several different ranges. The Blue Ridge Mountains begin in Pennsylvania near Gettysburg and end in Georgia and provide the breathtakingly beautiful backdrop for Asheville's culinary story.

The earliest inhabitants of what would become Asheville were Indians of the Mississippian culture, who settled in the area along the Swannanoa River near Biltmore Estate beginning around 800 to 1000 CE. The Mississippians are believed to be ancestors of several modern-day Indian nations, including the Cherokee.

From the beginning, these early inhabitants dined in an area of rich biodiversity said to rival the South American rain forest. The indigenous people who settled the Blue Ridge near what would become Asheville were skilled hunters, gatherers and farmers and would contribute much of their foods and foodways to what would come to be called Appalachian Mountain cooking. They lived peacefully (and alone) for several thousand years, but they were about to have company.

Sherrill's Inn was a popular dining spot for travelers on the Buncombe Turnpike in the early 1800s. *E.M. Ball Photographic Collection, D.H. Ramsey Library, Special Collections, University of North Carolina at Asheville.*

EUROPEANS ARRIVE IN THE APPALACHIANS

Spanish explorers led by Hernando de Soto were probably the first Europeans to see the Asheville area, visiting the Indian village at Jora, near present-day Morganton, in 1539. Although the Spanish had success in colonizing Florida and the western United States, this foray into the Appalachians was short-lived, and they abandoned their North Carolina forts after only a few years.

The next European visitors to North Carolina were English colonists at Roanoke Island in the late 1580s, the famous "Lost Colony." After this band of English settlers vanished, it would be another generation before the first successful English colony took hold at Jamestown, Virginia, in 1607.

The English chose to stick close to the coast for the next century, and it would be the mid-1700s before settlers from the United Kingdom arrived.

Beginning after the Battle of Culloden Moor, when the British crushed hopes for Scottish independence, Scots and the Scots-Irish (Scots who had colonized Northern Ireland) began pouring into the Piedmont and mountains of North Carolina.

During British colonial rule, settlements in the land of the Cherokee were forbidden by English law. But for Scots and the Scots-Irish, most of whom hated the British, the law was largely ignored. While no real settlements were established in the early years of the eighteenth century, many individual Scots made their way into the Blue Ridge.

As more and more settlers came into the area, the scenic valley made by the Swannanoa River as it joined with the French Broad River was a natural draw, and Asheville was founded in 1790.

The Buncombe Turnpike, following the French Broad from Greenville, South Carolina, to Greenville, Tennessee, was completed in 1828 and drew more settlers to the area. Asheville became a destination for people wanting to trade goods and deliver livestock to feed the growing population.

As more roads linked it to the rest of the state, Asheville began to get a reputation as a place to come for relaxation and recreation. When the railroad was completed in 1880, Asheville became a tourist destination.

Beginnings of Appalachian Mountain Cuisine

Climate, history and culture combined to create a unique blend of cuisines that would dominate the local food scene for the next two centuries and still tempts the taste buds of natives and tourists to this day.

Unlike the stereotypical vision of Indians, the Mississippians and later the Cherokee lived in cabins made from the abundance of trees in the primal forest that covered the Appalachians and Blue Ridge. Although skilled as hunters and gatherers, the Mississippians and Cherokee were also skilled at cultivation and agriculture.

Corn (maize), beans and squash were so vital to the Native Americans that they called them the "three sisters." Maize was often cracked by pounding in wooden blocks and boiling it to make what the Europeans would later call "grits."

Pumpkins, pecans, black walnuts, chestnuts, persimmons, blackberries and muscadines were also eaten by the local Indians. Deer, bear, rabbits, squirrels, possums, wild turkeys and groundhogs were also part of the earliest local cuisine.

The Spanish would have a huge impact on the cuisine of many nations as they explored North, South and Central America during the sixteenth century. As they explored, they took back examples of food from the Americas to Europe, touching off a culinary exchange between the Old and New Worlds that lasted for over two hundred years. Called the Columbian Exchange, this massive cross-pollination between Europe, Asia and the Americas is why we have Chinese oranges in Florida, South American tomatoes in Italy, chocolate from Mexico in Switzerland and hot peppers from the Caribbean in Thailand.

The Spanish were also responsible for bringing potatoes, vanilla, lima beans, sweet potatoes, peanuts, peaches and hot peppers from their original homes in Central and South America to Asheville tables.

When the Scots and Scots-Irish arrived on the scene, they befriended the Cherokee and adopted many aspects of their culinary traditions. The Cherokee taught the Scots-Irish how to smoke rainbow trout and how to make jerky out of deer meat. The Scots, who had no hang-ups about behaving like "proper English gentlemen," readily adapted to the diet of their native teachers. They introduced metal plows, Irish potatoes, turnips and the ubiquitous hog to the Cherokee, and a partnership began that lasted for years and resulted in a great number of intermarriages between the two groups.

European settlers also brought metal plows and began tilling and sometimes terracing the land, and soon the same crops that marked Southern cuisine in the eastern part of North Carolina began to spring up. These included onions, black-eyed peas, collards, turnips, sweet and Irish potatoes, pinto beans and watermelons. Apples, prized for their taste and "keeping" ability when dried, also became a favorite crop.

While the terrain wasn't suitable for large-scale cattle production, the pig was perfectly suited for the mountains and is still the star of many a meal in the area. Their ability to forage for themselves, ranging over the mountainsides eating whatever they could find, coupled with their prolific breeding ability and ability to produce more meat per pound of food intake than almost any other animal, made hogs a vital part of the diet of the early settlers.

The cuisine that developed from the melding of Scottish and Cherokee food traditions can still be found in the twenty-first century. Country ham, smoked trout, corn, pinto (or "soup beans" as the mountain folks call them), grits, corn bread, corn dodgers, field greens, turnip greens, poke "sallet" (salad), sourwood honey and dried apple stack cakes are dishes that would be as welcome on a table in 2013 as they were in 1813.

The average person living in the mountains of Western North Carolina would see few changes in his daily fare for nearly two centuries as isolation and a predominantly agrarian economy kept the wheels of culinary evolution turning rather slowly. For people in the most rural areas, the diet on a family subsistence farm in 1946 was basically the same as in 1846 and 1746. But for folks who lived in Asheville proper, change was on the horizon.

Chapter 2
DINING AT BILTMORE

Quite a few hotels in Asheville offered the finest food available, served on fine china flanked by highly polished silver, at the turn of the twentieth century. But for all their elegance, they paled when compared to the opulence of dining at Biltmore House.

Modeled after three sixteenth-century French chateaux, Biltmore was the home of George Washington Vanderbilt, one of the richest men in America. Designed by famed architect Richard Morris Hunt, it was completed in 1895. The house features four acres of floor space, 250 rooms, 35 bedrooms and 43 bathrooms, making it the largest private residence in America, a title it still holds today.

An invitation to dine with the Vanderbilts was a much sought-after prize among Asheville society. George and Edith Vanderbilt were known for lavish dinners centered on the finest foods and wines available.

Stocking the Larder

Putting food on the table at Biltmore was a grand undertaking that never stopped, even when the family was away at one of their other houses.

George Vanderbilt ensured that the estate was largely self-sufficient, thanks to its farmers, dairymen and herders, orchard men and gardeners.

The Banquet Hall at Biltmore Estate. *Photo courtesy of the Biltmore Company.*

The estate produced beef, lamb and mutton, pork (both domestic and wild boar), turkeys, chicken, mountain trout, milk, grains, honey, fruit and fresh vegetables for the Biltmore House tables.

Goods that weren't available on the estate were sourced locally, usually from merchants in Asheville's City Market, including the Asheville Fish Company and J.F. Miller, "purveyor of fish, oysters, clams, lobsters and scallops." Miller also offered squirrels in his list of products available. Other merchants included the French Bakery and Frank O'Donnell, "Importer, Jobber, and Dealer in Fine Brandies, Whiskies, Ale, Porter, Mineral Waters and Cigars."

There are also records of cheese obtained from a Mr. Yates, who ran a grocery on Haywood Street, and for spaghetti and macaroni from the Waldensian Bakery in Valdese, North Carolina. Some of the more exotic offerings at the Biltmore table, such as foie gras and truffles, came from merchants in New York.

Meats, dairy products, grains, fruits, game and vegetables, both from the estate and procured in Asheville and elsewhere, all came into the estate's main kitchen on the lower level via the kitchen courtyard, a busy place where wagons from the estate farms, orchards and dairy, as well as local grocers and meat markets, came to unload their goods. The food was taken into the kitchen, where it was carefully inventoried and stored in pantries, iceboxes, larders and root cellars.

The Kitchens

When Biltmore House was filled with the Vanderbilt family and their guests, more than fifty servants would have hustled in and out of the downstairs kitchens, pantries and servants' quarters.

There were three kitchens: the enormous main kitchen, where most of the meals for the family and house staff were prepared; a rotisserie kitchen off to the side of the main kitchen, where meat was roasted on electrically turned spits; and a pastry kitchen where baked goods, breads and desserts were prepared.

The main kitchen had a large coal-fired cookstove, a built-in charcoal grill and large hanging racks for copper pots and pans. There were also numerous gadgets to make serving meals for so many people a little easier, including a coffee grinder, sausage stuffer and giant mortar and pestle. Commercial-grade cast-iron storage cabinets held dry goods such as flour and sugar.

The rotisserie kitchen featured a large, wood-burning oven for roasting venison, turkey, chickens and wild boar taken from the estate's forests. The pastry kitchen boasted a special refrigerator to keep the proper temperature for chilling delicate pastry dough and a pastry oven for baking breads and desserts.

The main kitchen at Biltmore was well equipped and featured state-of-the-art appliances. *Photo courtesy of the Biltmore Company.*

With most midday meals and dinners requiring six or more courses, servants were always pulling china, flatware, glassware and silver from the first-floor butlers' pantry in preparation for setting the table in the Banquet Hall.

KITCHEN STAFF

The 1900 census showed that the kitchens at Biltmore were staffed by one chef, a thirty-eight-year-old Englishman named B. Vivian. He was assisted by James Ceperlean, a twenty-nine-year-old French cook.

There were twelve women employed in the estate kitchens in 1900, ranging in age from fifteen to fifty, and working as cooks and assistant cooks. Eleven of the women were from Western North Carolina.

The kitchen staff was under the direction of Emily King, the head housekeeper at Biltmore House. Mrs. King met with Mrs. Vanderbilt daily to plan and write out menus in a menu book.

One of these menu books, from 1904, was donated to the estate's archives in 2000 by a descendant of Esther Anderson, who was a cook at Biltmore that year. The book contains menus for luncheons and dinners served between September and December 31, 1904, along with comments on the meals written into the margins by George and Edith Vanderbilt. It has been an invaluable tool for helping the estate's curators learn about dining at Biltmore at the turn of the twentieth century.

Days started early for the kitchen staff at Biltmore. Even before the cooks arrived, scullery maids, who lived in quarters near the downstairs kitchen, would be up before dawn to light the fires in the ovens and cookstove and to prepare the pots and pans needed for breakfast.

The kitchen staff and servants had to be ready to attend to the family and guests at first light, so breakfast for the servants was an early affair. The kitchen staff would have eaten meals cooked in the main kitchen, consisting of mostly estate-grown meats, fruits, vegetables and grains, prepared more simply and closer to traditional Southern Appalachian fare. Personal servants to the Vanderbilts and their guests would have had meals more similar to those served to the family.

Dining at Biltmore

Meals for the Vanderbilts and their guests were served in the Breakfast Room or the Banquet Hall on the first floor. The massive oak table in the Banquet Hall is seven feet wide by twelve feet long when closed and extends to forty feet with all its leaves in place. Notes found in one of Mrs. Vanderbilt's

Biltmore's banquet table ready to receive guests. *Photo courtesy of the Biltmore Company.*

dinner books suggest that the largest number of dinner guests seated for one meal was thirty-six in April 1908.

Porcelain menu tablets were used for meals served in the Banquet Hall. Menus were written in grease pencil on these tablets and placed at each place setting so guests could see in advance what delicacies made up the next course.

Luncheon

While the Vanderbilt family often took breakfast in their rooms or sitting rooms, luncheon was the first formal meal of the day. In keeping with Edwardian tradition, luncheon was served precisely at 1:30 p.m. each day.

Luncheon was a lavish, multi-course affair, usually starting with fish, seafood or a soup course. Some of the family's favorite fish dishes, according to the surviving menu book, were codfish rolls with bacon, fish croquettes and creamed fish. Fried and scalloped oysters were also a favorite. The soup course at lunch was most often chicken broth or beef bullion.

The second luncheon course was almost invariably an egg dish. Favorites included egg cutlets (chopped boiled eggs mixed with a wine sauce to bind them together and then fried to resemble pork or veal cutlets), stuffed eggs or egg croquettes. Between breakfast and egg dishes for luncheons, invoices in the estate archive show the kitchen could regularly go through 115 dozen eggs in a month. Eggs were also combined with other ingredients for luncheon dishes such as sardines on toast with eggs.

The third course was meat, poultry or game, served with two or three side dishes of vegetables, macaroni or rice. Family favorites were roast chicken, roast duck with applesauce and lamb chops. Frequently served side dishes included potatoes, green beans, beets and rice.

Salad made up the fourth course of the proper luncheon, consisting of lettuce and other vegetables grown year-round in the greenhouses on the estate. Salads were usually served with an accompaniment of beef, lamb, pork, seafood, poultry or game, such as duck, rabbit or quail.

Luncheon desserts were usually light (no wonder, after all that other food), generally puddings and custards. Apple dishes that made use of the twenty varieties that grew in the estate's orchards were also popular with the Vanderbilt family.

After lunch, afternoon tea was served around 4:00 or 5:00 p.m., usually in the tapestry room.

Dinner Is Served

Dinner at Biltmore was an exercise in opulence. Dinner was served at 8:00 p.m., and both family and guests dressed formally for the meal.

The evening meal was more elaborate than lunch and usually entailed seven courses, although for special occasions such as Thanksgiving, Christmas or birthdays, the menu could stretch to eight or nine courses, or even more. The seven courses consisted of soup, fish, entrée, releve, salad, dessert and café noir. Black coffee with sugar was served as its own course at the end of the meal and was thought to aid digestion.

The Vanderbilts almost always had a house full of guests. Edith Vanderbilt's dinner book, which covers the period from November 1898 to April 1908, shows that dinner frequently involved between eighteen and twenty-one guests, but the number of guests could swell to as many as thirty-six.

As with lunch, the soup course at dinner was usually variations on consommé, a broth made from beef, poultry, vegetables or a combination thereof. Cream soups made from chicken, turkey, tomatoes, oxtail and mutton were also served.

The second course at dinner was some type of fish, often accompanied by a sauce containing eggs, cucumbers or tomatoes. Hollandaise or tartar sauce was also frequently served with fish. Bass, either caught on the estate or purchased from fishmongers in town, was in frequent rotation. Trout, red snapper, halibut and Spanish mackerel were some of the other fish mentioned in the 1904 menu book. On special occasions, oysters on the half shell were served as an appetizer before the soup course.

Copper molds were used to shape the third course, usually more elaborate dishes designed to showcase the culinary skills of the kitchen staff. Organ meats such as calves' brains and sweetbreads were popular for this course.

Entrées included turkey, ham, steak, mutton, duck, partridge and goose, which were prepared by roasting, frying, braising, broiling or fricasseeing.

The fourth course, called the releve, was what most twenty-first-century diners would consider a full meal, most often a roast or joint of meat or an entire bird accompanied by two or three starches and vegetable dishes. Cranberry sauce, horseradish sauce and mushroom sauce were served with the meats. Applesauce and fried hominy were served with roast duck, and cranberry sauce accompanied roast turkey, as it does on our Thanksgiving tables more than a century later.

Vegetables served with the releve course included asparagus, artichokes, beets, spinach, peas, string beans and sweet potatoes, many of them grown in the estate's gardens.

The fifth course was salad, similar to the ones served at lunch but usually more sophisticated and often including game birds such as squab, quail or partridge.

At this point, most twenty-first-century foodies would be groping around in pocket or purse for the antacid tablets, but the stout-hearted guests at the Vanderbilt table still had two more courses to go.

Dessert served at dinner was far more elaborate than at lunch. One of the Vanderbilts' favorites was Charlotte Russe, a dish that was extremely popular in the American South from the end of the Civil War until the 1920s. Nearly every period cookbook featured several variations on Charlotte Russe, which is made by lining a russe mold with ladyfingers and then filling the interior with a flavored Bavarian cream and chilling.

Desserts often consisted of cakes, cream puffs and éclairs, apple dumplings, pumpkin pie and brandied peaches, followed by a final course of black coffee.

THANKSGIVING AND CHRISTMAS DINNER

The dinner served to the Vanderbilts and their guests at Biltmore House on Thanksgiving Day 1904 included roast turkey with cranberry jelly, sweet potatoes, beets and peas. Guests also dined on oysters on the half shell, broiled Spanish mackerel, cutlets of calves' brains with mushroom sauce and a salade of roast Virginia ham with tomatoes and celery. Dessert was mince pie, a Thanksgiving favorite, served with pineapple ice cream.

According to the curatorial staff at Biltmore, Christmas luncheon in 1904 included broiled oysters, venison steak with string beans, potatoes and cauliflower and a salade with roast partridge. An apple tart was served for dessert. The venison almost certainly was harvested from the estate, likely by the estate rangers, whose responsibilities included providing wild game for the kitchens, including wild turkeys, grouse, partridge, quail and duck. Christmas dinner was almost identical to Thanksgiving with the exception of a traditional Christmas dessert of plum pudding. Plum pudding was another molded dessert made of beef suet, raisins, currants, citron and apples and flavored with brandy and spices—in other words, what is now sold as mincemeat. At the Vanderbilts' table, the pudding was served flamed with brandy and accompanied by sweetened brandy butter.

The Legacy of Biltmore

Although the Vanderbilts didn't invent fine dining, they did bring a whole new level of dining to Asheville. Many of the founding principles of dining at Biltmore are still present and practiced today in the estate's restaurants and bistros. Take, for example, the concept of locally grown, sustainable produce, much of it produced on the estate and picked within twenty-four hours of eating. That's the hallmark of today's farm-to-table synergy between local farmers and restaurateurs. Biltmore meals incorporated free-range chickens, beef and pork and organic farming techniques, combined with the vision of talented chefs. These founding culinary ideals continue to make dining at Biltmore Estate as special and memorable as it was when George and Edith Vanderbilt welcomed guests to their home.

Menu for Mr. Vanderbilt's Forty-Second Birthday

Blue Point Oysters on the Half Shell
Choice of Consommé Julienne or Puree de Volaille
Broiled Halibut with Sauce Hollandaise and Cucumber Salade
Chicken Mousse with Mushroom Sauce
Salade of Mutton with Currant Jelly, Rose Potatoes and Petit Peas
Pastry Cheese Puffs
Virginia Ham with Spinach
Pineapple Salade
Brandied Peaches with Vanilla Ice Cream and Assorted Cakes
Café Noir

Chapter 3
EARLY EATERIES

Asheville's earliest restaurants were in the grand hotels that sprang up after the railroad opened the town to tourist traffic in 1880. The Swannanoa Hotel opened that year, followed by the Battery Park in 1886. It was the hotel's commanding view of the Blue Ridge Mountains that inspired George W. Vanderbilt to purchase 145,000 acres of pristine mountain land for the site of his summer house.

The Battery Park was followed by the Langren in 1912, the Grove Park Inn in 1913 and the Kenilworth Inn in 1918. Patrons of the town's Grand Opera House could stop in after a performance and feast on steak from Texas or Maine lobster, both brought in by the new refrigerated railroad cars that used ice from frigid northern waters to keep the delicacies fresh on their way to Asheville.

One of the earliest freestanding restaurants was a Greek restaurant established about 1900 by Demosthenes Psychoyios, who also was known as "Barbathimo."

David Gross was another early restaurant pioneer. Born in Budapest, Hungary, in 1965, Gross came to America in 1873. Shortly after the turn of the twentieth century, Gross opened a cafe on Pack Square. When the Great Depression hit Asheville in the late 1920s, Gross was prepared with a sandwich stand downtown and a sandwich wagon on Broadway, all dispensing his famous hot dogs and ham sandwiches. Gross's sons, Leon and Charlie, went into the family business, and the restaurant was open for nearly a half century, closing in 1945.

The S&W Cafeteria. The building, which still stands, is considered a showcase of Art Deco architecture. *E.M. Ball Photographic Collection, D.H. Ramsey Library, Special Collections, University of North Carolina at Asheville.*

Another early Asheville eatery was Tingle's Café, owned by brothers Alvis Malburn (A.M.) and Thomas Tingle. The Tingle brothers moved to Asheville from Swansboro on the North Carolina coast and opened a downtown fruit stand before opening Tingle Brothers Café in 1918, located at 29 Broadway across the street from J.A. Keaton & Sons Lunchroom, which had opened three years earlier in 1915.

According to an oral history by Billie Walker Tingle (A.M.'s daughter-in-law) on file at the D.H. Ramsey Library at the University of North Carolina at Asheville, the café was famous for its coconut pie. "Chicken and dumplings was the best selling thing on the menu," Mrs. Tingle said.

In the 1930s, the restaurant went to twenty-four-hour-a-day service for about ten years but settled on "6:00 a.m. to 11:00 p.m. every day except Sunday," she said.

In the late 1940s, the Tingle family opened a second location called Tingle's 2 on Patton Avenue in West Asheville, which featured the same menu but was a drive-in, the hot new thing at the time. The drive-in was the first to close, and finally the downtown Tingle's closed sometime in the late 1970s. It had a resurgence in 2010, when a group of local investors brought back the café at the same location, but it closed in 2011.

In the 1920s, Guillet's Cafeteria served "Better Food" to hungry customers from its 21 Haywood Street location downtown. The decade also saw the birth of two legendary Asheville eateries, one of which was the Hot Shot Cafe at 7 Lodge Street in Biltmore Village. This all-night diner proved popular with railroad employees and residents of South Asheville. It opened in the mid-1920s and remained in operation under a succession of owners until it closed in 2005.

The other was the S&W Cafeteria, which opened its Art Deco–inspired doors at 60 Patton Avenue in 1929. Part of a chain founded by Charlotte businessmen Frank Sherril and Fred Webber, it was a popular spot for meetings and meals until it closed in 1974. The building, designed by architect Douglas Ellington, was placed on the National Registry of Historic Places in 1977 but remained vacant until 2007, when Steve Moberg purchased and renovated the building. It became home to the S&W Steak and Wine and a coffee shop called Corner House until they both closed in 2011.

"THE BLOCK": AFRICAN AMERICAN–OWNED RESTAURANTS

Asheville restaurants enjoyed great prosperity during the years between the end of the First World War and the stock market crash in October 1929. But in the days of segregation and Jim Crow laws, the city's African American population was barred from the downtown white-only restaurants. They found solace and good food at dozens of black-owned restaurants, cafés and other eating establishments on "The Block," the city's historically black business district.

Bordered by Market and Eagle Streets, "The Block" was home to dozens of restaurants. In 1920, there were fifteen "Eating Houses, Restaurants and Cafes" listed in the city records, including: R.D. Alexander's, 10 Eagle Street; B.F. Brown's at 99 Valley; the Bull City Cafe at 380 Depot Street; Charles Charter's place at 160 Southside Avenue; G.F. Hamilton's at 26 Eagle Street; and Samuel Johnson's at 382 Southside. Clara Morris owned an "Eating House" at 89 Eagle, just down from W.H. Owens at 78 Eagle. Bud Payne's Eating House was located at 415 Ralph Street, H.P. Pearson had his place at 6 Eagle Street and the Rex Cafe served its customers at 448 Depot. A.W. Williams was down the street at 386 Depot, and Hattie Giles ran her place at 8 Butrick.

In 1892, George Vanderbilt constructed a community center on The Block for the African American craftsmen who helped build Biltmore Estate. Called the Young Men's Institute, the building at 39 South Market Street housed a number of businesses over the years, and by 1920, it housed the YMI Cafe.

Building Back after the Depression

The 1930s brought the F.W. Woolworth chain of five-and-dimes to downtown Asheville with the opening of its store at 25 Haywood Street. The building featured a popular soda fountain that was a fixture for generations of Asheville teenagers. The soda fountain and lunch counter was the site for

Interior of Perkin's Cafe, 1938 or 1939. *E.M. Ball Photographic Collection, D.H. Ramsey Library, Special Collections, University of North Carolina at Asheville.*

several sit-ins during the 1960s as African Americans in Asheville joined the movement to desegregate lunch counters across the South.

Although Woolworth's is long gone, the soda fountain is alive and well, serving up sandwiches, floats, shakes and malts inside Woolworth Walk, the city's largest indoor art gallery.

Another iconic Asheville eatery born during the Depression was the famous Sky Club. Located in a majestic three-story mansion atop Beaucatcher Mountain, the Sky Club was originally named the Old Heidleberg after owner Gus Adler's hometown in Germany. After World War II started, Gus changed the name to the Sky Club to avoid anti-German sentiment.

The Sky Club survived Gus's death in a fire in 1952 and continued until the building was sold by his wife and children in 1975.

The Sky Club was known for elegant dining, excellent New York strip steaks and shrimp cocktail and for spectacular views of downtown.

Minority-owned restaurants continued to boom in the 1930s, many of them characterized as "lunchrooms." They included Albert's Place at 86 Eagle; George Black's lunchroom at 386 Depot Street, in the spot formerly occupied by A.W. Williams; the Boston Cafe at 393 Southside Avenue; J.W. Butts's lunchroom at 221 Southside; Josephine Darty's lunchroom at 118 Vance Street; and Eagle Lunch, open for business at 127–29 Eagle Street. Hamilton's Café was located at 26 Eagle, J.A. Hill had a lunchroom at 6 Eagle and Carrie McMillan ran her lunch place at 106 Eagle. On South Market, patrons could choose from the Elk's Cafe at 29, the Lincoln Cafe at 26, the Toggery Grill at 31 and the Wayside Inn at 46 South Market. Southside Avenue boasted a tearoom at the Booker T. Washington Hotel at 411; the Worthy Cafe at 189; C.N. Nueble's lunchroom at 406; and Hines Moore's place at 209 Southside. Mattie Harris owned a lunch place at 157 South College, while Lillie Johnson tempted customers with the smells of home cooking at 462 South French Broad and Pearl Mango did the same at 10 Morrow. Edith McGowan had a place on Riverside Drive, Millie Samuel was at 144 Poplar and Nora Shelton greeted customers at her 190 Blanton Street cafe. The S&J Tea Room at 3 Wilson Street and the Quality Cafe at 43 Buttrick rounded out the list.

The 1940s brought the Silver Dollar on Clingman Avenue, open to serve breakfast, lunch and dinner to folks who worked the mills, warehouses and industries that once occupied the space along the French Broad River. For years, it served as an industrial zone, but the area is now transformed, occupied by potters, weavers and glass blowers who call the River Arts District home.

But back to the Silver Dollar—it even made it to the silver screen as a location for a scene in the 1958 moonshine-running classic *Thunder Road*. When the Clingman Avenue bridge was built, the diner was moved to a new home on higher ground, which saved the building from a disastrous flood in 2004 that took out several restaurants near the river. The restaurant served burgers, hot cakes and Greek favorites under the direction of Catherine and Angelo Dotsikas, who bought the restaurant from the original owners in 1966. They finally closed the doors in 2011.

War-Years Eateries

By 1942, black-owned restaurants had exploded, and Eagle Street was the place to go for any type of food. You could start your walking feast at Alice Robinson's place at 12 Eagle, stroll a few doors down to the Palace Grill at 19 and then hit the Silver Star Café at 25. The Delretta Sandwich Shop was next door at 26 and vied with the Harlem Sandwich Shop at 34 Eagle for the lunch crowd. The names of the establishments were exotic, if nothing else.

There was the Elite Café at 57, Minnie's Place at 69 Eagle, Albert's Place at 81 Eagle and Effie Hentz's cafe at 88 Eagle.

The Ever-Ready Café rounded out the list at 107 Eagle, and the Rumbe ice cream shop kept everyone cool with cones and cups at 67 Eagle Street.

On Southside Avenue, the restaurants started with Silver Moon Sovoria at 103, Blue Ridge Barbecue at 221 and Leo's Place at 334 Southside. M&W Café was at 393 Southside, with Marion's Ice Cream Parlor at 187 and Midget Café at 209 Southside.

South Market Street boasted the Downtown Café at 24½, Sallie Taylor's at 42 and Williams Restaurant at 46 South Market.

Mountain Street had the East End Tavern at 44 and the El Parvo Cave at 37. Other black-owned restaurants of the 1940s included the Glamorest Tea Room at 178 Blanton Street; Blanton Street Grill at 83 Blanton; Casaloma Beer & Lunches at 101 Choctaw; Darling Tea Room at 147 Poplar; Stanford Hudson's place at 23½ Crescent; Jake's Chicken Shack at 22 Palmer; Madden's Tea Room at 6 Congress; and Mother's Kitchen at 28 Beaumont Street.

POSTWAR EATERIES: DINERS AND DRIVE-INS

When the GIs returned to Asheville after the Second World War in the late 1940s, they brought a new perspective to eating in the mountains.

Guys who had been raised on hog and hominy came back with a taste for pasta, pizza and Chinese food. Banks were eager to loan money to start new businesses, and everyone wanted a new car after wartime production had previously made them impossible for civilians to own. New highways were being built, and drive-in restaurants meant you didn't even have to get out of your car to enjoy a burger.

Asheville's first drive-in was Buck's Restaurant, started in 1946 by John O. "Buck" Buchanan. The sprawling restaurant, the parking lot of which covered nine acres at the site of the current Applebee's on Tunnel Road, was an Asheville icon for three decades.

At its peak, Buck's employed 165 people, according to an oral history recorded by Buchanan in 1994. While teenagers circled the parking lot and pulled in for burgers and malts, their parents flocked to the Red Carpet Room, the fanciest of the restaurant's six dining rooms. It seated one hundred diners at a time and had a larger and more sophisticated menu than the drive-in.

Buck's Restaurant was a popular Asheville eatery. *Photo courtesy the University of North Carolina Libraries.*

The second of the "Big Three" Tunnel Road eateries in the '50s and '60s was Wink's, owned by Bill Winkenwerder. As teenagers came out of the tunnel on Tunnel Road, Wink's was the first stop of the cruising ritual. Smaller than Buck's, Wink's had a powerful draw in addition to its Wink burgers, Dagwood sandwiches and fries. In a stroke of marketing genius, Winkenwerder put a radio tower on top of the drive-in and bought all the radio time from 8:00 p.m. until midnight six nights a week on WISE radio. A disc jockey perched high atop the tower would lower a peach basket on a rope down to would-be Romeos, who would send up their requests for songs to be dedicated over the air to their sweeties below. In an interview from the early 1990s, Winkenwerder said the first night the DJ was on duty, word spread like lightning among area teenagers, and a traffic jam ensued that backed up from the site of the current Papas & Beer on Tunnel Road all the way through the tunnel and onto Charlotte Street.

On the other side of Tunnel Road was Babe Malloy's, owned by George C. Malloy. After completing their orbits of Buck's and Wink's, teenagers of the '50s and '60s turned left out of Buck's and pulled into Babe's, which featured shaved ham sandwiches and legendary chocolate shakes.

The last year of the 1950s saw the founding of an Asheville landmark that would dish out Greek and Italian food to hungry diners for more than half a century.

In 1959, the Zourzoukis family started a diner where the Montford Bridge over I-240 stands today called the Express. After the bridge came, they relocated to 183 Haywood Street, where they changed to the Greek and Italian formula that would win them numerous "Best of Asheville" awards in those categories.

The family renamed the restaurant Three Brothers following the move to Haywood Street, after the three Zourzoukis brothers who owned the restaurant. They stuck with the name when a fourth brother joined the team, and the formula kept Three Brothers popular until it closed in 2011 after a fifty-two-year run.

In 1946, Erline McQueen opened an Asheville icon on "The Block," at the corner of Eagle and Market Streets: the Ritz. McQueen chose the name to give her restaurant "a little class," she reported in a 1995 oral history interview preserved at the University of North Carolina–Asheville library. After her first attempt to purchase a space ran into difficulty when her partners backed out, McQueen moved into the old Masonic Temple in 1951 and began a restaurant dynasty that lasted for the next thirty-two years.

McQueen devoted herself to her restaurant, working most days from 6:00 a.m. to 11:00 p.m., frying chicken and fish, making pork chops and pot roast and cooking vegetables bought fresh each day from the farmers' market on Lexington Avenue or obtained from her cousin's farm. Each morning, the workingmen would stop by for the special: two sausage biscuits and coffee for thirty-five cents.

The Ritz was more than just a good place to eat; in segregated Asheville, it was a favorite gathering place for African American business owners to meet and make deals over McQueen's delicious home cooking. Many a business deal was sealed by a handshake over a piece of sweet potato pie and a cup of coffee.

Roast beef, steaks, ribs and French fries were always on the menu, every day except Tuesday, McQueen's only day off. The Ritz was a favorite place for Sunday dinner until McQueen finally decided that "the Lord has helped us so much we needed to go to church" and started closing on Sunday instead of Tuesday. She died in 2005 at the age of ninety-five, a respected business and civic leader.

THE 1960S AND BEYOND

The later years of the 1960s and the 1970s brought quite a few favorites to the Asheville dining scene.

Little Pigs BBQ, founded by Joseph Carr Swicegood Sr., began serving pit-cooked pork in 1963 and is still going strong.

Barbecue Inn, with its iconic "Blinky Pig" sign overlooking Patton Avenue, served pork barbecue and Brunswick stew to Ashevillians from 1961 until the owners closed it in July 2011.

Another favorite Greek/Italian restaurant from the period was Athenian Bistro, located on Biltmore Avenue near McCormick Field. It opened in 1968 and lasted until March 2013. Another 1960s favorite, Pete Apostolopoulos's the Mediterranean Restaurant on College Street, has the honor of being the oldest restaurant in downtown Asheville. It has been serving breakfast, Greek specialties, burgers, seafood and sandwiches since 1967.

Chapter 4

THE FARMERS

Farming, while challenging in any terrain, was a necessity with the settlers who crossed into the Blue Ridge Mountains. But it was a hardscrabble fight against gravity to get it going.

Just think of trying to farm in the mountains. Horace Kephart, in his book *Our Southern Highlanders* (1913), tells of "a Kentucky farmer who fell out of his own cornfield and broke his neck. I have seen fields in Carolina where this might occur, as where a 45-degree slope is tilled to the brink of a precipice. A woman told me: 'I've hoed corn many a time on my knees—yes, I have.' And another: 'Many's the hill o' corn I've propped up with a rock to keep it from fallin' down-hill.'"

Bottomlands, flat flood-prone areas near rivers, proved to be fertile fields much easier to till and tend. Still, early farming was small and marginal.

Yet long before today's top chefs discovered free-range, organic and sustainably grown foods, they were the main ingredients found in the stewpots and on the tables of the earliest settlers to the area.

Early agriculture in the mountains surrounding Asheville was mostly subsistence based. The main items produced on nineteenth-century farms were corn, beans, squash, potatoes (both Irish and sweet) and apples. While some farms kept a cow for milk and butter and a few chickens to provide eggs (and an occasional chicken dinner when they became too old to lay), pork was the main meat produced, for a variety of reasons.

Hogs were prolific breeders, having as many as a dozen piglets each spring. They were the epitome of free range: they wandered the woods eating pretty

Biltmore Dairy Farms truck in the estate courtyard, circa 1900. *E.M. Ball Photographic Collection, D.H. Ramsey Library, Special Collections, University of North Carolina at Asheville.*

much anything they could find. By winter, when the temperature dropped below freezing the first time, it was "hog-killing time," and all the piggies except for the breeding stock were turned into sausage, bacon and salt-cured ham for the year ahead.

Asheville, writes Wilma Dykeman in *The French Broad*, "was a little island of self sufficiency."

While the main goal of early farmers was always subsistence, any surplus crops were a readily available source of cash or barter. Corn not bound for the gristmill or destined for overwinter feeding of the breeding hogs was turned into corn liquor, a much easier commodity to transport and trade. Apples, either whole, dried or pressed into cider, were also popular. And most Appalachian farm wives had a secret hiding place for the closest thing there was to a bank account in those days: the "butter and egg" money carefully saved from sales to neighbors or the cooks from the railroad gangs or logging crews.

FARM TO FORK, 1890S STYLE

By the later years of the nineteenth century, a movement had begun to apply scientific methods and land management to farming. Some of the movement's earliest practitioners were the farmers George Vanderbilt hired at Biltmore Estate.

From the beginning, the farming operations at Biltmore were tasked with providing meat, milk, poultry, vegetables and other farm commodities to the table at Biltmore House, providing income to the estate through the sale of farm products and serving as a sort of laboratory for demonstrating successful farming techniques to farmers and educators.

Baron Eugene d'Allinges was hired to be the first head of the estate's farming operation, which was called Biltmore Farms and also encompassed Biltmore Dairy. A native of Saxony, Germany, d'Allinges had immigrated to the United States and was practicing scientific farming techniques in nearby Skyland, North Carolina. He was soon discovered by Vanderbilt's farming consultant, Edward Burnet, and was enlisted to farm at the estate in 1889 or 1890. By 1897, Biltmore Farms letterhead listed "A.J.C.C. [American Jersey

Ox team pulling a sled at Hickory Nut Gap Farm, date unknown. *E.M. Ball Photographic Collection, D.H. Ramsey Library, Special Collections, University of North Carolina at Asheville.*

Cattle Club] Jerseys and Butter, Southdowns [sheep], Berkshires [hogs], Prize Winning Poultry, Market Garden Products, [and] Sourwood Honey" among the bounty produced for the Biltmore House table and the local market.

After d'Allinges died in 1895, George F. Weston took over the operations. According to an 1896 edition of the *Asheville News and Hotel Reporter*, "Mr. Weston…is a fine chemist and graduate of an agricultural college and a farmer of some twelve or fourteen years experience."

In May 1897, Biltmore Farms placed an ad in the *Asheville Citizen* offering dressed broilers at thirty cents each, table eggs at fifteen cents a dozen and "dated and extra selected" eggs at twenty cents a dozen. By the end of that same year, enough eggs were being produced to supply all the needs of Biltmore House as well as an additional eight to fifteen dozen a day to supply the Kenilworth Inn.

The Market Garden was an important part of supplying the table at Biltmore. In addition to forcing houses (an early name for greenhouses), it also boasted several greenhouses for the production of lettuce, a highly sought-after and hard-to-grow ingredient for the salad courses served at luncheon and dinner on the estate.

The soil in the Market Garden was enriched organically with manure as well as fertilizers. It took a huge amount of manure to enrich the fields, much of it coming from the cattle operation. The latest scientific farming techniques were employed by George Stevenson Arthur, a Scotsman who was hired to manage the Truck Farm and greenhouses.

Vegetables grown at Biltmore's Market Garden included Irish potatoes, queen sweet potatoes, reedland early drumhead cabbage, white plume celery, yellow aberdeen turnips, purple top strap leaf turnips, purple top rutabagas, cow-horn turnips, parsley, rapier, dwarf Yosemite mammoth beans, telegraph cucumbers, garden broad leaved cress, fine curled cress, scarlet turnip forcing radish, silver skin onions, Boston market lettuce, champion sweet corn, crown prince peas, sibley squash, new stone tomatoes and long orange carrots. Fruits included new early Hackensack muskmelons and grapes.

MODERN FARM TO TABLE AT BILTMORE

Today, about 2,500 acres of Biltmore Estate is devoted to farming, which includes a Market Garden, the vineyards, fields for grains and forages and pastures for cattle, sheep and horses.

A herd of about two hundred registered Angus cattle are kept pastured on the estate's rolling hills, of which thirty to fifty steers each year supply the grass- and grain-fed beef for the estate's five restaurants. A flock of White Dorper sheep, numbering more than 150 ewes, was added in the fall of 1996. More than 100 lambs are harvested each year for use in the Biltmore lamb program.

The Field to Table Production Garden supplies vegetables, herbs and small fruit year-round, all of which are used in estate restaurants. Everything from common home garden items such as lettuces, heirloom tomatoes, asparagus and winter squash to more unusual fare such as red scallions, five different colors of carrots, edible flowers and micro greens find their way to the estate's tables.

Seasonal vegetables not produced on the estate come from various farms in surrounding counties, in addition to several small mountain farms that are part of the Appalachian Sustainable Agriculture Project. Apples and apple cider are selected from orchards in Hendersonville.

Mountain Farms (Almost) Go Up in Smoke

After the Second World War, farmers in the Blue Ridge were faced with a new choice. As more and more family farms were abandoned after the war and commercial farming grew in the Midwest and California, more and more meat and produce began finding its way into area grocery stores, which sprang up in record numbers. Since subsistence farming was no longer a necessity, mountain farmers began to devote more and more acreage to cash crops, especially Bright Leaf and Burley tobacco to feed North Carolina's tobacco industry. This led to good times for local farmers, as season after season of good tobacco crops led to less and less land devoted to crops.

By the late 1980s, the small farms that dotted the bottomlands and hung onto the hillsides of the Blue Ridge were under attack. Tobacco had fallen victim to falling prices and mounting public pressure as cigarette companies paid out huge settlements in class-action lawsuits. Meanwhile, stiff competition from larger farms in the eastern part of the state had prices down and many local farmers wondering if it was time to sell out to the land developers circling their farms like buzzards.

CHARLIE JACKSON AND THE APPALACHIAN SUSTAINABLE AGRICULTURE PROJECT

Charlie Jackson and the Appalachian Sustainable Agriculture Project were pioneers in the sustainable, community-supported agriculture movement. *Amy Kalyn Sims.*

In the 1990s, Charlie Jackson was one of those small farmers eking out a living on a piece of property he owned up in Madison County. Charlie wasn't your typical mountain farmer, though.

He grew up in Durham, about as far away from the mountains as you can get without hitting saltwater. But after earning a degree in environmental history from Appalachian State University, the mountains became his home. After a brief stint at the University of Maine to earn a master's degree in environmental history, Jackson moved to Madison County and started farming.

His academic background led him to find a grant from the Kellogg Foundation in the mid-1990s to examine and develop local solutions to the farming crisis. Jackson wrote a proposal, and Asheville was one of the communities selected.

"Tobacco was on its way out, and we needed to find some ways for mountain farms to stay viable," Jackson said. "We were losing a lot of farms at that time. Tobacco had been the foundation for the economy of many of our rural communities, and it was going away."

For the first few years, Jackson and a few others, all volunteers, brainstormed, planned and held community forums to find a way to make the small mountain farms survive, if not thrive.

Around 1999 to 2000, Jackson and his volunteers had come up with the idea of local food as a way to increase the value of local farms.

"It was only thirteen or fourteen years ago, but at the time this was pretty radical stuff," Jackson said. "Today, it's the biggest food movement in the world."

Jackson began to get the word out about the superiority of local food and the superior taste of the heirloom fruits and vegetables grown for generations by mountain farmers.

With no money for his own salary, let alone advertising, Jackson turned to guerrilla marketing a decade before the term was invented. "One of the best things I ever did was to buy a digital camera," he said. "I started taking pictures of produce at the markets, and food is easy to make look beautiful."

Armed with a list of farmers' markets and tailgate markets across the mountains, Jackson approached Polly McDaniel, then features editor at the *Asheville Citizen-Times*, about running a list of markets and what was in season. He found a ready ally in McDaniel, a master gardener who grew up on an Indiana farm.

"We began a professional relationship that continues to this day," Jackson said. "We supplied articles, photos and market schedules, and Polly ran them in the paper. It resonated with people, and it became an important part of getting the word out."

As people began to get out to the markets and interact with farmers, they were introduced to new and different varieties of fruits and vegetables only found on local farms. The difference in a tomato bred for maximum flavor versus one bred to have a skin tough enough to survive a cross-country journey made quick converts of local cooks.

In 2002, Jackson and crew—who had by then gained 501 (c)(3) nonprofit status under the name Appalachian Sustainable Agriculture Project (ASAP)—began publishing a local food guide, which listed everything from you-pick farms and farm stands to tailgate markets and restaurants featuring locally sourced foods.

The guide was an immediate success, required reading for everyone from natural food–loving hippies and hog farmers to executive chefs and tourists. From five thousand copies the first year to its current press run of one hundred thousand copies a year, more than one million copies of the guide have been distributed since its inception.

As ASAP began to assist farmers with replacing tobacco with heirloom vegetables, Jackson realized that most of the tobacco farmers didn't know how to market the new crops. With tobacco, they brought the year's crop down the mountain all at once, made one stop at the auction house and that was it for the year. So ASAP started the Business of Farming school to teach marketing, bookkeeping and other business skills.

To make sure "local" remained a meaningful term, in 2006 ASAP started the Appalachian Grown program, a branding and certification program that currently boasts sixty farms and three hundred businesses. "It's our tool to ensure the value of local stays with our farmers, particularly now that the word is such an attractive marketing tool," Jackson explained.

Today, ASAP has three main programs: the Local Food campaign, which includes the food guide and Get Local, a promotional campaign where restaurants feature a local product each month; Growing Minds, which provides locally grown foods to schools through the Farm to School program; and the Local Food Research Station, which studies the crops that can benefit local farmers.

From its grassroots, all-volunteer origins, ASAP has grown to a staff of twenty-one and is a nationally recognized leader in the farm-to-table and farm-to-school movements. What was once a small group of volunteers making it up as they went along is now teachers who promote model programs followed nationwide by other organizations with similar goals.

THE FARMERS

The farmers responsible for Asheville's arrival on the national food scene make up a diverse group, some newcomers to the land, some multi-generational farmers. Some are college graduates, and some were more than happy to put their high school in the rearview mirror. Some planned on being farmers from childhood, while others just seem to have stumbled upon it. But all of them share a passion and love for the land and the crops they grow.

Like their forebears at the turn of the twentieth century, they face cold winters, late frosts and early springs that make plants bloom before the last frost has reared its head. But the farmer at the dawn of the twenty-first century faces challenges from global climate change, encroachment from developers, stacks of government paperwork and competition from huge agribusiness conglomerates determined to replace heirloom vegetables with genetically modified superplants grown for their convenience rather than their flavor.

In spite of the challenges, these farmers pull into the area's many tailgate markets each week and deliver the raw materials that make our restaurants some of the finest in the country.

Jamie Ager
Hickory Nut Gap Farm

Jamie Ager feels a strong connection to the land he farms, one that has run for nearly a century through the Hickory Nut Gorge. Five generations of descendants of Presbyterian minister Jim McClure and his wife, Elizabeth, have farmed the beautiful, rugged land that the couple bought in 1916.

The property features a rambling old stagecoach stop called Sherrill's Inn, which stood alongside the Buncombe Turnpike (also called the Drover's Road). The turnpike served as the main thoroughfare from the farms that dotted the coves and bottomland of the Southern Appalachians and the markets in the Piedmont of North and South Carolina. Cattle, hogs and other livestock would be herded down the turnpike to market, with the inn a popular resting place for man and beast.

Jamie, the McClures' great-grandson, and his wife, Amy, have been running Hickory Nut Gap Farm since 2000, using sustainable and organic farming methods to preserve the land that has been in the family for the last century.

Both Agers are graduates of Warren Wilson College in Swannanoa, a school with a deep farming tradition that teaches sustainability by having its students work on the college farm to grow food for the school's dining hall.

Pastured pigs at Hickory Nut Gap Farm. *Amy Kalyn Sims.*

After graduation, the couple put what they had learned to work on their own farm and continued to study farming methods by traveling to other countries during the slow winter months while the farm was dormant. So far they have studied farming systems in New Zealand, Hawaii, Italy, California, Brazil and Argentina.

Hickory Nut Gap is renowned for its pastured pork, grass-fed beef and poultry, which is sold in local grocery stores as Hickory Nut Gap Meats. The farm also supplies more than twenty-five area restaurants with bacon, sausage, beef, pork and chicken. In addition, ten acres of certified organic apples, blackberries, raspberries, black raspberries and asparagus are also produced on the farm.

In the spring of 2009, a permanent conservation easement was placed on the farm's 350 acres of working farmland. This easement will preserve the integrity of Hickory Nut Gap Farm as a working farm for all generations to come.

Annie and Isaiah Perkinson
Flying Cloud Farm

Just down the road from Hickory Nut Gap Farm, Jamie Ager's cousin Annie Louise Perkinson and her husband, Isaiah, manage Flying Cloud Farm, an organic vegetable, berry and flower operation also located on the McClure/Clark legacy farm.

Annie (the family calls her Annie Louise to differentiate her from her aunt Annie) grew up in the shadow of Hickory Nut Gap, her grandparents' farm. Although her parents weren't part of the farming at Hickory Nut Gap (her father, an Englishman, was the community's country doctor), she still spent her childhood working the land.

"I remember canning, making jam and freezing the summer bounty with my grandmother and mother so we could have delicious food to eat all year," she said. "We lived an 'agricultural lifestyle,' growing and preserving much of our own food, drinking raw milk from my grandparents' dairy and working when there was work to be done.

"As the oldest of five kids and twenty-one cousins who grew up in a 'back to the land' household on a family farm, I wanted nothing to do with kids, animals or dirt when I left home at eighteen," she recalls, laughing now.

Annie Louise left the farm after she graduated from high school and set out for the University of North Carolina at Chapel Hill, where she met

Annie Louise and Isaiah Perkinson, managers of Flying Cloud Farm. *Amy Kalyn Sims.*

Isaiah while both were working at a Franklin Street pizza place. After two years at UNC, she took a year off and traveled in Europe, working on two farms, an organic operation in Germany and a farm in England that had a community-supported agriculture program (CSA), where community members bought an advance stake in the farm's production.

"These experiences opened my mind to the different kinds of agriculture around the world and made me see potential for our family farm in Fairview," she said.

After graduating from Warren Wilson College, Annie Louise and Isaiah decided to make a go of it and rented land from family members.

"Isaiah and I decided that we wanted to love our work, live and raise our kids together and grow food to feed our local community," she recounts. "We got started in 1999 by planting a field of blueberries and a half-acre garden. We started our CSA in 2002, with both of us working full time on the farm during the busy growing season."

The Perkinsons worked toward being full-time farmers for almost ten years. "Isaiah is a skilled carpenter and master mechanic, which are essential to running a successful farm. I was part owner of Trout Lily Market, a small natural foods store in Fairview. By 2007, we were both on the farm year-round."

Flying Cloud Farm grows fourteen acres of organic vegetables, berries and flowers. The breakdown is roughly one acre of strawberries; one acre of blueberries; half an acre of blackberries; one and a half acres of sweet potatoes; one acre of potatoes; one acre of winter squash; one acre each of peppers, eggplant, squash and tomatoes; two acres of brassicas; and several more acres of mixed vegetables.

Annie Louise and Isaiah started selling to area tailgate markets from the earliest days of Flying Cloud, and some of their earliest and most frequent customers were chefs. The same is true today.

"We primarily sell to the chefs that come to the market to buy food for their restaurants," Annie Louise said. "These are mainly Zambra's, Southern, Tod's Tasties, Chocolate Lounge, West End Bakery, Local Joint and Rosetta's."

To Annie Louise, Asheville's food scene and farm-to-table movement is just more folks discovering what she grew up knowing. "I grew up eating homemade food, raw milk, grass-fed beef, homemade jam and chicken we processed ourselves. When I left home, I deviated from this awesome diet and realized how important good food is for health and general well-being."

Sally Eason
Sunburst Trout Farms

From the earliest days of the Cherokee, the people who settled these mountains knew the sweet, earthy, succulent taste of rainbow trout. But it took a Yale engineering dropout to get the rest of the world interested.

In 1948, Dick Jennings returned to the Blue Ridge Mountains he had loved as a child, where he spent carefree summers tramping his grandfather's farm and fishing for trout in the cold, clear waters that raced down from the high peaks. On the site of his grandfather's farm, he established the first commercial trout farm in the South, selling the beautiful, multicolored rainbows to stock streams and rivers to keep fishermen happy.

With the new interest in eating fish spawned by the heart disease prevention research of the 1960s, Jennings began processing and selling trout for its meat, and by the 1980s, he renamed the Jennings Trout Farm to Sunburst Trout Company to cover the addition of his daughter Sally and her husband, Steve Eason (it's since changed again to Sunburst Trout Farms).

Sally Eason of Sunburst Trout Farms, a pioneer in sustainable aquaculture. *Amy Kalyn Sims.*

Some of the early customers at Sunburst included famed Southern restaurant Aunt Fannie's Cabin in Atlanta; Giant Foods in Washington, D.C.; and the Washington Fish Exchange in D.C.

In 1986, tired of having to dispose of hundreds upon hundreds of pounds of roe (fish eggs) during spawning season, Sunburst set out to see what could be done with the bright orange eggs. After several trial-and-error recipes and negotiations with a buyer at New York's Caviarteria, the caviar from Canton was soon the talk of New York chefs.

Today, Sunburst is on its third generation with the addition of Sally and Steve's sons, and nearly every local restaurant and supermarket sells some of the sixteen different trout products it harvests from the cold waters that flow down the mountain from the Shining Rock Wilderness.

Sally, who runs the farm today, loves the synergy that happens when local chefs use local products. "Chefs can explain who and where the farmer is, and the product is ultra fresh," she said. "You have innovative chefs integrating local ingredients into their cuisines, and the result is grassroots gone viral in Asheville."

Bruce DeGroot and Karen Mickler
Yellow Branch Farm

When Bruce and Karen saw the beautiful mountain cove in the Yellow Branch community in Graham County near Fontana Lake, they knew they'd found home.

Bruce, a carpenter and furniture maker, and Karen, a potter, were passionate gardeners who wanted a place where they could grow their own food. The fifty acres they bought in 1980 along with family members fit the bill nicely. They had no idea that they were about to start an organic cheese-making operation that would supply restaurants from Asheville to Atlanta.

Karen and Bruce bought their first cow in 1981, and a few years (and cows) later, they had shifted from growing organic vegetables for local restaurants to making cheese.

Karen attended cheese-making classes in Canada and Wisconsin to refine her skills and developed the recipe that they use today. Yellow Branch cheeses are known for their buttery, mild flavor, which Bruce and Karen attribute to the grasses that make up the diet of the cows. The pastures are managed using sustainable organic practices with no herbicides, pesticides or chemical fertilizers, and growth hormones are not given to the cows.

Above: Bruce DeGroot and Karen Mickler supply cheese to many top Asheville chefs and restaurants. *Amy Kalyn Sims.*

Right: Bruce DeGroot of Yellow Branch Farm making cheese. *Amy Kalyn Sims.*

Yellow Branch makes several varieties of cheeses, including Yellow Branch Farmstead Cheese, a mild, buttery, full-bodied table cheese; Yellow Branch Pepper Cheese, made with organically grown jalapeño peppers from Karen and Bruce's garden; Yellow Branch Tomato Basil Cheese, which is made every fall using garden-raised basil and dried heirloom tomatoes, also from the garden; and Yellow Branch Natural Rind Cheese, which is unwaxed and washed with organic apple cider vinegar. It is aged for a minimum of four months, and the result is a drier, sharper version of the Farmstead Cheese.

Bruce and Karen started working early on with some of Asheville's top chefs, and you'll find Yellow Branch cheeses on the menus of Early Girl Eatery, HomeGrown Restaurant, Laurey's Catering, the Biltmore Coffee Shop, Sunny Point Cafe and restaurants on Biltmore Estate.

Walter Harrill
Imladris Farm

Growing up to be a farmer was about the furthest thing from Walter Harrill's mind when he was a kid. His parents had been the first generation in their families to leave the farm and attend college. Other than a few weekend visits and some time during the summers, the family farm didn't fit into Walter's idea of happily ever after.

A degree in zoology with a minor in genetics led him to a career as a medical technologist, where microscopes and test tubes were a lot more common than tractors and pickup trucks.

After ten years of working in hospitals and labs, Walter and his wife, Wendy, also a medical technologist, got a chance to move rent-free into a vacant house on his grandfather's farm in Fairview (eastern Buncombe County). They jumped at the chance.

Walter's grandfather had a pick-your-own blueberry farm, and at the time (the late 1990s) Walter had just become aware of what organic was. He knew that his grandfather had used organic methods all his life and that the blueberries he was selling dirt-cheap were worth much more. "I knew Grandpa was working very hard and getting into his eighties, so I told him I'd come help," Walter said. "I also knew he should be making a lot more money for the amount of work he was putting into the crop."

After talking with Annie Louise Perkinson at Flying Cloud Farm, Walter took the blueberries to the local tailgate market. After selling fifteen pounds in twenty minutes the first week and forty pounds in half an hour the second week, Walter realized he was on to something. He also realized he was running out of time.

"At this point, we had about three weeks of blueberry season left and about three months of tailgate season left," Walter said. He went home and brainstormed with Wendy, who by this time had quit her job. Wendy suggested they try making jam. "That was the dumbest idea I ever heard," Walter replied. But Wendy insisted, and Walter found himself up to his elbows in blueberry jam.

After selling jam at the tailgate market, Walter became one of the first vendors contracted for the outdoor market at the Grove Arcade. One day a young chef named John Stehling walked by, tasted the jam and asked Walter to make him a sales pitch to supply jam to his new Early Girl Eatery.

The relationship was good for Walter, who by now had christened the jam and the farm "Imladris," the Elven name for the village of Rivendell in the *Lord of the Rings* trilogy. Stehling gave feedback, and Walter tweaked recipes and wholesale packaging.

For the first three years, Walter and Wendy made their jam right in the farmhouse kitchen. As demand grew, it became a little more than the small space would bear. "The last year we did ten thousand jars in the kitchen of our house," Walter said. "It burned out a stove and almost burned up a marriage," he quipped.

Today, Imladris has seven acres of berries that go into its jams, which come from a gleaming commercial kitchen that poses no danger to home stoves or marital bliss. Imladris jams are sold in about twenty-five restaurants from Charlotte to Charleston and in several gourmet markets.

All this has led to a happy couple who prefer jelly jars to test tubes.

Chris Sawyer and Missy Huger
Jake's Farm

It shouldn't come as much of a surprise in a community where we have a chef who's a neuroscientist and a brewer who's a nuclear engineer that we would have an organic farm run by a former stockbroker and a carpenter.

Chris Sawyer and Missy Huger own Jake's Farm, a certified organic farm located on Beaverdam Creek near Candler, a few miles west of Asheville. Chris grew up near Beaumont, Texas, dreaming of being a farmer and spending happy boyhood days on his uncle's farm on the Brazos River. But a hard dose of reality caused him to put his farming dreams on the backburner.

"I was in seventh grade when I realized I wasn't going to inherit any land and probably wasn't going to be rich enough to buy any," Chris said. "So I became a carpenter."

For Missy, born and raised in Asheville, farming was the furthest thing from her mind. "The closest I ever got to farming as a child was visits to my friend Susie Clarke Hamilton's home, Hickory Nut Gap Farm," she said. "My focus was always on travel."

After spending two years living in Switzerland as a child, Missy returned to Asheville and attended University of North Carolina–Asheville before moving to Savannah, Georgia, and attending Armstrong Junior College and then taking a job at Merrill Lynch.

Wanderlust led her to move to England, where she worked for a couple American brokerage firms, ran an office for a Saudi Arabian plastics company and private art dealers and even worked for Warren Beatty while he was in England making the feature film *REDS*.

Back in the States, things weren't going very well for Chris. In 1995, while working as a self-employed ceiling contractor, he fell from a scaffold. Although he was only six feet off the ground, the fall broke both arms and led to emergency back surgery a week later to correct a pinched nerve affecting bladder function.

"Two months later, the disc above ruptured, and I was crippled for over six months before back surgery number two happened."

In the early 1980s, Missy moved to Los Angeles, where she worked as vice-president of member services for American Film Marketing Association, which ran an international trade show called the American Film Market. This kept her busy with trips to the Cannes Film Festival for fifteen years, along with numerous other business trips to Europe.

In 1998, she returned to Asheville, where she had met Chris in 1992 while he was doing some work on her mother's house. "Imagine my surprise when, in late August of '98, Chris appeared at my house and announced he had quit his job. 'And what are you planning on doing?' asked I. He replied, 'We are going to farm.'"

Chris bought almost eleven acres of rough land, secured an ancient backhoe and started shaping and contouring the land. They brought in their first crop in 1998 and were certified organic in 1999.

"We joined Carolina Organic Growers in 1999, which was a wonderful experience for us," Missy said. "We were very fortunate at the time there were several people involved who were kind and generous enough to mentor us. We were also fortunate to be invited to be on the ASAP farm tours for a number of years. Finally, we were really too busy and getting too old to dress the farm up for that event."

Missy and Chris started working with local chefs early in their farming career. "I started working with local chefs, most notably, Mark Rosenstein [formerly] of the Market Place," Missy said. "Mark was the first in this area to really be interested in local and sustainable. I very much enjoy working with Jacob Sessoms of Table, Brian Canipelli of Cucina 24, Duane Fernandes of Isa's Bistro, Reza of Rezaz, William Dissen of the Market Place and Michael Moore of Seven Sows among others."

For anyone thinking of following in their footsteps, Missy emphasizes that farming is hard work. "Chris spent untold hours studying about dirt, compost, fertilizer, amendments and, most importantly, bugs," Missy said. "There is absolutely nothing romantic about farming. It's hot, dirty, wet or too dry."

Chapter 5
PIONEERING CHEFS

Asheville's dynamic dining scene owes a lot of its drive to a group of pioneering chefs who took to heart the adage that "a chef's day begins in the garden."

That's true to the point that it sometimes brings foraged food to the table. Asheville chefs quickly recognized that they had come to a town that was not only a little off the beaten path but also a little off kilter when it came to food.

Take John and Julie Stehling of Early Girl Eatery, for example. Cooking came second to finding fresh food. "We spent the first year we were living here going out to the farms and meeting the farmers," John said. "We wanted to have that kind of connection to our food."

Called by some the "San Francisco of the South" for its thriving counterculture, Asheville was and still is home to a large population of people who eat vegetarian, vegan or organically (if not all three).

Any of the farmers you met in the previous chapter will tell you that they owe a lot to the chefs in this chapter, just as these chefs will gladly tell you that without the farmers, their cuisine would be a shadow of what it became with their help.

The chefs you'll meet here are nationally recognized as pioneers in Asheville's locally sourced food movement. They are also nationally recognized as forming the farm-to-table blueprint for the rest of the country to follow during the past decade.

Although they came to Asheville from different backgrounds and from different parts of the country, they all brought with them a passion for excellent food made with the freshest ingredients from farm and field.

Mark Rosenstein
Founder, the Market Place

Ask anyone about the genesis of fine restaurants in Asheville, or the start of the nationwide farm-to-table movement, and you'll always hear Mark Rosenstein mentioned as one of the first to grasp the notion.

Rosenstein, who grew up in Broward County, Florida, was a student at Northwestern University in Evanston, Illinois, heading toward a career in radio and television when a wine dinner and a wayward Frisbee changed his life. It also ended up changing Asheville dining forever.

"I had been to this wine dinner giver by a Belgian merchant my freshman year, and the food and wine was incredible," Mark said. "After that, I knew I wanted to do something with food and wine."

Mark Rosenstein has been called "The Godfather" of the farm-to-table movement and is a nationally recognized authority on sustainable practices for restaurants. *Amy Kalyn Sims.*

In 1972, after he climbed a large oak tree to retrieve a Frisbee while on Christmas break, Rosenstein unknowingly got a severe case of poison ivy from a vine snaking its way up the tree. "I started back to school, and I was itching and swelling like crazy. About Highlands, North Carolina, my eyes were swelling shut, so that was as far as I got."

In Highlands, Mark found an interesting and eclectic community of people, many of them into organic farming and foraging for wild foods.

"As a kid in elementary school, I had an adult friend who was a Cherokee Indian and an honorary Seminole chief. He connected me to the natural world, foraging in the woods and the seashore for turtle eggs and plants you could eat—that was a big influence on me."

After poring over books on food and wine, his new passions, Rosenstein abandoned the notion of returning to school and set out to build a restaurant from the ground up.

"My friends and I built the restaurant ourselves," Mark said. "We did all the carpentry, plumbing and wiring ourselves."

Rosenstein began looking for local farmers and foragers to provide the ingredients for his new restaurant, dubbed the Frog and Owl Café, out of a combination of idealism and necessity.

"I was reading Escoffier and Louis Diat, learning classic French cuisine, and they said a chef's day began in the garden, selecting his ingredients," Mark said. "The basic tenet I picked up was that a chef has a relationship with his garden."

He was helped along by a slightly less philosophical reason for local sourcing: "We were located five miles up a one-lane dirt road, and none of the food service delivery companies would send their trucks up to us."

"I started buying produce from Ruby Harper up on Gold Mine Road," Mark said. "She had the best farm in the area. I was going to Atlanta on Monday when we were closed to get produce at the Atlanta Farmers Market and getting some meat from Asheville as well. When we would run low on trout, I was bringing buckets of live trout through the dining room because we hadn't built the back stairs yet."

Never missing an opportunity to learn about food, Mark trained with French cooking teacher/cookbook author Simone Beck and Madeleine Kamman, founder of the School for American Chefs. Other training came from Rosenstein's work in the kitchen of Roger Verge at the Michelin-starred Moulin de Mougins in the south of France.

Before long, the little restaurant at the end of the dirt road had attracted the attention of *Gourmet* magazine, as well as rave reviews from customers all over the country who came to Highlands' resorts and discovered the local gem.

In 1979, Mark came to Asheville and started scouting space for a fine-dining restaurant downtown. At the time, downtown Asheville was the poster child for urban blight. The Pack Square area was home to pool halls and porno shops. Fine dining was chicken wings from a bar, and vintage wine was anything older than last night, usually served from a brown paper sack.

He found space on Market Street, and the Market Place was born. Two years later, he moved the critically acclaimed restaurant to Wall Street, where it remains a destination restaurant for Asheville foodies.

Rosenstein continued to develop relationships with local farms, farmers and food groups. He has served on the board of ASAP and is a longtime member of Slow Food Asheville.

In 2010, he sold the Market Place to chef William Dissen and embarked on a new career as a culinary educator. His current passion is Green Opportunities, where he is program manager for GO Kitchen Ready, a program that trains low-income adults with barriers to employment for jobs in the restaurant industry.

After more than forty years in the kitchen, Rosenstein is more passionate about food and local sourcing than ever. He finds it "hilariously ironic" that people are making such a fuss over the "revolutionary big thing" he's been doing all along—sitting down to eat locally sourced food, cooked slowly and lovingly to share with family and friends.

VIJAY SHASTRI
FOUNDER, FLYING FROG CAFÉ
MR. FROG'S SOUL AND CREOLE KITCHEN

In the mid-1980s, one of the hottest restaurants in town was the Windmill European Grill, which featured an eclectic blend of dishes from all over Europe. The restaurant was the first of several groundbreaking Asheville eateries owned by the Shastri family.

Just a year or so after its opening, buzz was circulating about the family's son, Vijay, who had worked his way up from the cold pantry to a place on the line and was demonstrating amazing culinary insight for a teenager.

Vijay, who spent his early childhood in India, grew up in a multi-ethnic family where food was always center stage. "The aromatic spices of Indian cuisine, slow-braised German dishes and sweets were everywhere in our home," Vijay says.

He soon developed a reputation for using masterful combinations of spices to flavor dishes across a wide variety of ethnic cuisines. By 1989, he and his sister struck out on their own to open Café Bombay, which featured a number of his favorite flavors from childhood. It was the first of many executive chef positions Vijay would occupy.

By 1994, Vijay was owner and executive chef at the Latin Quarter, a total shift in focus and cuisine.

The following year, Vijay launched his most successful venture to date, the iconic Flying Frog Café and its companion bistro, the Frog Bar. The Frog, as most locals referred to it, was a Mecca for fine dining at a time before Asheville had many high-caliber dining spots.

The Flying Frog's menu was an eclectic world tour of flavor built around produce sourced from the local farmers' market and area farms, featuring French, Italian, German, Indian and even Cajun favorites, all masterfully prepared. The restaurant also allowed Vijay to apply another of his passions: fine wines. The wine list at the Frog was among the finest in the state, according to longtime patrons.

Several years in a sluggish economy signaled the end of the Frog in 2011. The following year saw Mr. Frog's Soul and Creole Kitchen, which had a short May-to-December run before the building it occupied on "The Block" in the historic African American business district was taken over for renovation into a multi-use space featuring retail and affordable housing. As this is written, Vijay is scouting for space for his next Asheville restaurant.

Vijay embraced the farm-to-table movement that sprang up among local chefs.

"I've always sourced as much local product as possible," Vijay says. "My preference is towards local produce, eggs and cheeses. I have used regularly Cane Creek Organics, Camp Celo Farms, Cloud Nine, Spinning Spider Creamery and several locals at the farmers' market," he says.

"As local farmers are using better practices and feed programs for their livestock, I'll be supporting them on that end as well," he added. "I try as much as possible to use regional livestock as I can. Quality is still the most important part of the food selection process."

JOHN AND JULIE STEHLING
FOUNDERS, EARLY GIRL EATERY

Fresh vegetables from local farms, free-range and pastured meats and Southern cooking like Sunday dinner at Grandma's house have always been the hallmarks of Early Girl Eatery, John and Julie Stehling's farm-to-table restaurant in downtown Asheville. Even before the restaurant opened, the couple made a commitment to supporting local farms and giving their customers the best and freshest food possible.

The couple, early supporters of Asheville's burgeoning farm-to-table culture, came from widely different backgrounds.

"I grew up just outside Winston-Salem, and for years my father was a baker at Old Salem," John says. "My parents always had a big garden, and fresh food we grew ourselves was always a part of my life."

After getting a degree in restaurant management from East Carolina University, he went to Charleston to help his brother Robert in the kitchen of his critically acclaimed restaurant, Hominy Grill. There he met front-of-the-house manager Julie, who grew up in inner-city Detroit. Soon they began planning the move to Asheville.

"My parents were into backpacking and kayaking," John relates, "and we spent a lot of summers in Asheville. I've known for fifteen years that I wanted to have my family here."

After moving to Asheville in 2000, John went to work in the kitchen at Savoy and Julie went to the Golden Horn (now closed). They both worked at Tupelo Honey Café when it opened, all the while looking for space for a restaurant of their own.

Looking for a space and scrimping every penny they could, the couple was between rentals, using the library for an office and living in a tent while waiting for the lease on an apartment and a restaurant space to go through. After a trip to Hot Springs Resort to celebrate their wedding anniversary, they returned to Asheville at midnight to find their tent gone. "We had one set of friends in town at the time, and they thought it was hilarious that we had to come knocking on their door on our anniversary asking to crash at their house because our tent was stolen," Julie said.

A space came open on Wall Street when a Cajun restaurant closed, and Early Girl Eatery opened October 13, 2011.

One of Early Girl's hallmarks is its commitment to using fresh and organic produce as much as possible. John grew up helping his parents grow organic vegetables at their home in Kernersville, North Carolina, and even before the restaurant opened, the couple began seeking out local growers.

"We sought out the Appalachian Sustainable Agriculture Project," Julie said. "The first farm we started working with was Yellow Branch up near Robbinsville. We found them through ASAP."

Some of the earliest farms they worked with were Green Toe out in Celo, Hickory Nut Gap, Yellow Branch and Imladris, John said. "There are some others who were seasonal, but those were the core," Julie added.

"People from California are amazed," Julie said. "They come out here and find out we're serving local eggs and local sausage and local meat, and out there only the most high-end restaurants are able to use local ingredients because the stuff's just not there locally. And that just doesn't happen by accident—people are working to make that happen."

The hardest decision in the beginning was what to name the new restaurant. The couple e-mailed possible names to all their friends and found that "everyone loved and hated all the names equally," according to Julie. After rejecting suggestions from well-meaning family and friends for names such as J.J.'s and Evil John's, they turned their attention to the produce aisle. "We wanted something heirloom like squash or tomatoes," said John. "We eventually found Early Girl (tomato), and the logo just sort of popped into our heads."

In the twelve years since it opened, Early Girl's farm-based Southern cuisine has drawn praise from *Bon Appetit*, *Southern Living*, the *New York Times* and *USA Today*, as well as rave reviews from thousands of happy tourists and locals.

"Food matters here," Julie said. "Food always holds a place in a Southerner's heart."

BRIAN SONOSKUS
EXECUTIVE CHEF
TUPELO HONEY CAFÉ

When it comes to understanding the farm-to-table movement, there's no one more qualified than Brian Sonoskus. A New Jersey native, Brian grew up visiting grandparents who were active in community farms, where entire communities shared farmland and reaped the benefits of homegrown produce.

"My grandmother would handle the cows," Brian remembered. "She was the one who distributed the milk to the other members of the community."

Meals with his grandparents made an impression on Brian. "We were always together, cooking as a family," he said. "The kitchen was the center of the home. My grandmother had a couch and a TV in her kitchen, and it was where she spent her days. One of my earliest memories is making pierogies with her."

Becoming a chef was a natural progression for Sonoskus. "The only jobs I've ever had for money were in restaurants," he said. "I started out as a dishwasher at a seafood restaurant on the beach in Jersey, and within a few weeks I was boiling lobsters, then I was dropping hush puppies, frying shrimp and working on the line."

His high school guidance counselor suggested culinary school after graduation, so Sonoskus packed up and headed to Providence, Rhode Island, to the prestigious Johnson and Wales University, where he earned his degree in culinary arts.

He moved to Asheville in 1995 and started working at Magnolia's Raw Bar & Grille on Walnut Street.

In 2000, he happened into Tupelo Honey Café, a brand-new restaurant specializing in Southern comfort food. Sharon Shott, the owner at the time, was trying to run the restaurant and manage the kitchen at the same time, a daunting task for anyone. They talked, and Sonoskus has been there ever since.

Brian Sonoskus, executive chef at Tupelo Honey Restaurants, is a passionate gardener and former part-time farmer. *Amy Kalyn Sims.*

Brian began sourcing local produce as much as possible, hitting the Western North Carolina Farmers Market and the few tailgate markets available at the time. Soon, farmers heard about the restaurant and started contacting Brian about buying what they had picked that day. It was farm to table before anyone had even coined the phrase.

Then Sonoskus took it one step further and began growing food for the restaurant. From 2006 until 2010, Sonoskus had his own farm in Madison County, where he grew produce for his family as well as some for the restaurant. "I wanted my kids to grow up in that type of environment, and I wanted to grow everything," he said. But after current owner Steve Frabitore opened a second Tupelo Honey location in 2010, the long hours spent commuting from Madison County were too much. Brian sold the farm to a friend.

Brian's seasonal menus and creative specials have won him praise from *Southern Living* and the *New York Times*, and today he sources as much local food as possible, even bending the seasons a bit. "Ramps, for example, have a very short window," he said. "I'll buy them when they're in season and put them in the freezer. That way when it's winter and I want ramps, I can have some."

Brian uses Hickory Nut Gap Farm for his meats and advocates local sourcing of beef, pork, chicken and eggs. His reasoning: "Can you imagine the carbon footprint of a cow coming all the way from New Zealand?"

As Tupelo Honey has expanded (the company has locations in Greenville, South Carolina, and Knoxville, Tennessee, with plans to open in Charlotte, North Carolina), Sonoskus has noted that even significantly larger cities lack the culinary sophistication that Asheville has achieved.

"The number of restaurants, the quality of restaurants and the quality of employees we have in Asheville is amazing," he observed. "The consumer drives a lot of what we do, but the farmers definitely have a great impact on the cuisine."

Brian is committed to getting the word out about fresh foods and sustainability, working with First Lady Michelle Obama's Chefs Move to Schools program in elementary schools and underserved women and children. "I'll go into a school and lecture a little bit, and then we'll make something fresh, like a green salad," he said. The result? "You see kids eating that and enjoying something they never thought they'd eat. It really makes me feel good."

Laurey Masterton
Founder, Laurey's Catering

Laurey Masterton has been an icon on the Asheville food scene since she opened Laurey's Catering in 1987. She is the youngest of the three daughters of Elsie and John Masterton, founders of the original Blueberry Hill Inn in Goshen, Vermont, and authors of the three *Blueberry Hill* cookbooks.

After a childhood spent watching her mother create fresh, locally sourced dishes, Laurey earned a degree in theater lighting from the University of New Hampshire and tried careers in the New York City theater scene. She then became an Outward Bound instructor before settling down to cater to Asheville's hungry crowds. Laurey quickly became known for innovative cuisine made from fresh, local and organic ingredients.

"I started buying locally from the North Asheville Tailgate market when I started working in 1987. And when I moved downtown in 1990, I started

Laurey Masterton of Laurey's Catering. *Amy Kalyn Sims.*

working directly with Barry and Laura Rubenstein of B+L Organics," Laurey said. "I've been committed to buying from local farmers since the beginning of my business, buying directly, at the markets and through our local distributors."

Largely self-taught as a chef, Laurey interned for Nora Pouillon, the creator of the first 100 percent certified organic restaurant in the United States. She has been an integral part of the local farm-to-table initiatives, with a particular interest in helping children experience gardening, cooking and the eating of "real food." She's also a former board member of the Appalachian Sustainable Agriculture Project.

In 2010, Laurey was one of a coalition of chefs invited to the White House by First Lady Michelle Obama to discuss involving chefs in schools.

One of her favorite ingredients is local honey, and she's also a passionate fan of bees. "They pollinate one-third of all the food we eat," she says. She keeps bees herself. As a tribute to the hard work of her honey-making friends, Laurey wears a tiny vial around her neck containing one-twelfth of a teaspoon of honey, the lifetime production of a single honeybee.

In addition to leading culinary tours to Italy and France, Laurey has authored two books. Her first, *Elsie's Biscuits*, was a culinary memoir of growing up in her mother's kitchen and the recipes she remembered from her childhood. Her latest is *Fresh Honey Cookbook: 84 Recipes from a Beekeeper's Kitchen*.

Over the past twenty years, Laurey has catered concerts and film locations and has fed celebrities including Ricki Lee Jones, Aaron Neville, Dar Williams, Joan Baez, Leo Kotke, Garrison Keillor and Bob Dylan.

Laurey works with between fifteen and thirty local farms to provide the fresh ingredients for her creations and is passionate about protecting and conserving the land. "We have been recycling since the beginning too, collecting and sending our compostable materials to local farms or local gardeners," she said.

Her take on what makes Asheville unique for food lovers? "ASAP has been enormously important, making the farmers' and tailgate markets prevalent and available, which, in turn, has offered the opportunity for more people to produce local food," she said. "When I first moved here there was ONE tailgate market—behind the Fresh Market in North Asheville. Now there are many options for people wanting to buy local produce, making it much easier for all of us to fill our tables with really good ingredients."

Joe Scully and Kevin Westmoreland
The Corner Kitchen
Chestnut

One of Asheville's most successful restaurant partnerships started out in an unlikely place—a campfire for people not old enough to play with matches.

Kevin Westmoreland is a rare bird: an Asheville native. With Asheville a tweener between a cosmopolitan city and a mountain hamlet, it's becoming rare to find someone who's actually from here. "I went to T.C. Roberson High, and I live in the house I grew up in," Westmoreland said. "My grandmother made biscuits three times a day, and that's the type of food I grew up with."

After graduating from UNCA, Kevin embarked on a career as an information technology specialist with healthcare systems, ending up in Denver, Colorado, where he met his wife. They eventually ended up back in Asheville, drawn by the lure of family and the place he grew up.

When Westmoreland's son was five, he took him to a meeting of the Indian Guides, a YMCA program that emphasizes Native American culture and good values. There he met Joe Scully, a chef who grew up in New Jersey and was cooking at various restaurants around town. Several years of camping with their sons led to a lasting friendship and, unbeknownst to Kevin, a career change.

"About this time [January 2004], my company was sold, and I was faced with moving to Cleveland or not having a job," Kevin said. "Nothing against Cleveland, but we decided to stay here."

Scully, who had moved to Asheville after stints as executive chef at Cherokee Town & Country Club in Atlanta, several Manhattan restaurants and even at the United Nations, was working at Chelsea's Tea Room in Biltmore Village (now closed). He soon met Bill Hathaway, owner of a longtime coffee shop and restaurant right across the street.

"Joe told me he knew this guy who wanted to retire and wanted to know if I'd be interested in opening a restaurant with him," Kevin said. "I didn't know anything about restaurants, but I read everything I could get my hands on and talked to friends who had worked at or owned restaurants, and I decided to go for it. We took over the building in January and opened Corner Kitchen February 16."

Westmoreland's customer service skills combined with Scully's intuitive ability to cook what the pair dubbed "CaroAmerican comfort food" made

Chef Joe Scully and Kevin Westmoreland, founders and owners of the Corner Kitchen and Chestnut. *Amy Kalyn Sims.*

Corner Kitchen an immediate hit. Even President Barack Obama and First Lady Michelle dined there during a visit to Asheville.

Westmoreland and Scully began sourcing local foods even before the paint was dry on the Corner Kitchen's walls.

"What impressed me when I moved to Asheville was that it was this island in the middle of all this agriculture," Scully said. "I was doing a luncheon for the French Broad Garden Club, and there was this one woman who was different from all the rest," Joe said. "I asked who she was, and they told me, 'Oh, that's Annie Ager, she's the egg girl—she used to sell eggs in Biltmore Forest from her parents' farm.'"

Scully visited Ager at Hickory Nut Gap Farm, which her family had owned for generations, and the pair became fast friends. For Scully, it proved to also be a valuable farming connection. "I rode my bike out to the farm and met Jamie Ager and the rest of the family," he recalls.

"Thirteen or fourteen years ago [about 2000], farm to table was just incubating. I had read Alice Waters and the relationship she had with the farmers out in California, so I was well versed in the food industry, but doing something in town…I said, who else can I talk to?"

One connection led to another, and soon Joe had relationships with many of the pioneering farms involved with farm to table. The fresh ingredients and his creative presentations drew notice from national magazines, including three features in *Southern Living* in just eighteen months.

In September 2012, the pair continued their winning combination of farm fresh ingredients and creative preparation with Chestnut, their newest restaurant, on Biltmore Avenue in downtown Asheville.

"Farm to table in Asheville is the norm now," Kevin said. "If you're not using local foods, you're not in the game. In addition to the farms we work with, we grow our own herbs in the kitchen garden at Corner Kitchen and Joe grows some at his house," Kevin said.

Another trend Scully and Westmoreland are onboard with is the concept of green restaurants. "We compost and recycle everything we possibly can," Kevin said. "It used to be that food came in one way, people ate most of it and the rest went in the landfill. Now we've closed that loop."

DAMIEN CAVICCHI
CORPORATE EXECUTIVE CHEF
BILTMORE ESTATE

Biltmore Estate corporate executive chef Damien Cavicchi oversees Biltmore Estate's dining and culinary operations. *Amy Kalyn Sims.*

When most chefs want to visit the farms that provide the ingredients they use in their culinary creations, they have to drive to the country. Damien Cavicchi just has to step out the back door. Damien is corporate executive chef at Biltmore Estate, in charge of food and beverage operations at all the estate's restaurants. With Biltmore's commitment to sustainability, vegetables, fruit, beef and mutton come to Chef Damien's tables directly from the estate's farms and pastures, in addition to local farms.

Cavicchi had a childhood most chefs would die for. He grew up in an Italian family with a mother who cooked everything from scratch, surrounded by the Creole-French, Southern and Cajun flavors of his hometown of New Orleans.

"Dinner every night with my family was something we looked forward to," Damien said. "We had a big family, so it was usually entertaining, and my mom is an excellent cook."

Damien's father was also a culinary influence, even after his family "had the misfortune to move to Kentucky" when he was in high school.

"My dad wasn't a great cook, but his excitement for food was infectious," Damien said. "We would pick wild blackberries together and eat them with milk and honey. He'd get excited for crawfish and shrimp season, and we usually bought ours from the guys on the roadside who sold them out of their pickup trucks.

"Later, in western Kentucky, Dad would take us to smoky dives that had incredible Kentucky-style barbecue and to restaurants that gave me my first taste of country Southern food—which was different than the Southern food we'd had in New Orleans. He was always after the best food he could find."

All this exposure to food led to his first job at age fifteen as a busboy and dishwasher at a restaurant in Paducah, Kentucky. After graduating from high school, Cavicchi's family moved to Asheville, and he decided to attend culinary school at Johnson and Wales in Providence, Rhode Island. "I was attracted to the unique mix of self-discipline, intensity and creativity and hard work required to succeed as a chef," he said.

The structured regime of the famous chefs' academy wasn't for him, and he dropped out after four months. He then embarked on a cross-country journey, working at restaurants and studying under different chefs while learning the ropes in kitchens serving different cuisines.

The hard work paid off in a happy accident when the wife of one of the sous chefs he worked with took over the kitchen at another restaurant. Things didn't work out, and Cavicchi took over the kitchen. "I learned a lot from my cooks and by reading everything I could get my hands on," he said. "But the job was crazy—breakfast, lunch and dinner every day. I went for three months without a day off."

Damien worked his way back to Asheville and began a four-year stint as executive chef at La Caterina Trattori, where his use of local, seasonal produce earned him a reputation as one of Asheville's hottest new chefs.

In 2005, he opened his first restaurant, Clingman Café, in Asheville's River Arts District. A few years later, he sold Clingman Café to open Sugo Fine Food, an upscale modern Italian restaurant on Patton Avenue, downtown.

Damien joined Biltmore in 2009 and sees Asheville's development into a food destination as a fusion of the place, the chefs and the farmers.

"Even before Asheville was so recognized nationally, there was always an audience here for good food," he said. "Asheville is like Austin and Portland in some ways—a vibrant liberal arts scene seems to lend itself to a vibrant, albeit casualized, food culture. Chefs and farmers have great relationships here, I think due to the strong entrepreneurial spirit that people here seem to have."

CATHY CLEARY AND KRISTA STEARNS WEST END BAKERY

The old adage goes in part, "for want of a horse a kingdom was lost." In the case of West End Bakery, it could have said, "for want of a sandwich a bakery was born."

West End Bakery is a West Asheville icon, a constantly humming gathering place where you can find neighbors doing everything from discussing beekeeping to making business deals over a sandwich or a cinnamon roll the size of a VW hubcap. But there was a time when it was a vacant storefront, waiting for the right dose of tender, loving care.

Cathy Cleary was a cook for a summer camp in Brevard while she was in high school. "I was hired first as a farmer, but since I was passionate about cooking, I ended up in the kitchen," she said. "The connection between the garden and the kitchen has always been a strong one for me."

After graduating from the University of North Carolina–Greensboro with a degree in art history, Cathy moved to Asheville. She was working as a teacher at Rainbow Mountain School when she noticed something about her West Asheville environs. "As I walked to work each day, I noticed that

West End Bakery's Cathy Cleary (left), Lewis Lankford (center) and Krista Stearns (right). *Amy Kalyn Sims.*

there was nowhere for parents to get a healthy lunch for their kids," she said. "I was talking to my friend Krista [Stearns], and we were sad that there was nowhere for us to hang out besides our houses. We got to thinking that it would be great if there was somewhere besides the library for the community to gather.

"Krista and I started dreaming and talking, and the more we talked about it, the more we thought it could actually happen."

Armed with a small loan from a credit union, the pair enlisted husbands, friends and family to do plumbing, wiring, floors—anything necessary to turn an empty storefront into a hip neighborhood hangout.

"From the beginning, we wanted to use as many fresh, local and organic ingredients as possible, and before we even opened our husbands had their gardens in place with all kinds of lettuces and greens and herbs," Cathy said. "They even had their own business plan to start an urban farm, but they ended up starting the West Asheville Tailgate Market instead."

The couples also worked to establish the Haywood Market, a food co-op that lasted several years before closing down.

"Reid [Chapman], my husband and Lewis [Lankford], Krista's husband, worked with ASAP to start the West Asheville Tailgate Market," Cathy said. "They had all their meetings at the bakery, and I met the farmers we have relationships with. Tom Elmore was one of the early farmers we worked with, and Tom Sherry, who used to have Whistle Pig Farm, was also an early supplier. It's great to be able to buy produce from your friends."

As word spread and the bakery became a fixture, other farmers stopped by, many with farms too small to have names. "We have a loyal customer, Keith Byrum, who grows shiitake mushrooms. He comes in once a month, and we give him French bread for mushrooms," Cathy said.

In the past few years, Cathy has become really involved with the North Carolina Organic Bread Flour Project, which extends that relationship with the farmer to the miller and in that way gets feedback on the grains.

Both Cathy and Krista are actively involved in the community: Krista is heavily involved with local schools, and Cathy is currently president of the board for FEAST Asheville, an organization affiliated with Slow Food Asheville dedicated to promoting healthy eating in all socioeconomic groups.

"FEAST teaches cooking classes to children to get them cooking healthy food," she said. "I was a co-founder, and the bakery also donates heavily to FEAST. We're also a community meeting place for FEAST and other organizations."

The Table Is Set

The chefs in this chapter, along with a few other pioneers such as Hector Diaz of Salsa's and Zambra and Reza Setayesh of Rezaz, brought Asheville to the attention of the country's top food media. By the early 2000s, the sleepy little tourist town was beginning to get noticed by folks from all over the United States for the farm-driven, creative cuisine coming from its restaurants, coffee shops and cafés.

Chapter 6
FRESH

Asheville's passionate chefs and cooks have a variety of choices for sourcing fresh ingredients. At the turn of the twentieth century, there were local butcher shops, farmers' markets and farm wagons selling fresh-that-day produce.

Today, the tradition continues with a whole-animal butchery, the WNC Farmers Market, dozens of tailgate markets, farm stores and community-supported agriculture shares (CSAs).

The Chop Shop

Whole-animal retail butcher shops were pretty common at the turn of the twentieth century, but after the Second World War, commercially available meat in supermarkets soon sent them the way of the dodo. Now the concept has resurfaced in Asheville at the Chop Shop Butchery, 100 Charlotte Street.

The Chop Shop specializes in whole animals rather than the more common primal cuts (essentially gutted and quartered animals). It's the only full-service, whole-animal butchery in the state as of 2013.

Chop Shop owner Josh Wright says whole-animal butchery is more sustainable than using primal cuts because it promotes using the entire animal without waste.

Chop Shop Butchery owner Josh Wright and head butcher Tyler Cook. *Amy Kalyn Sims.*

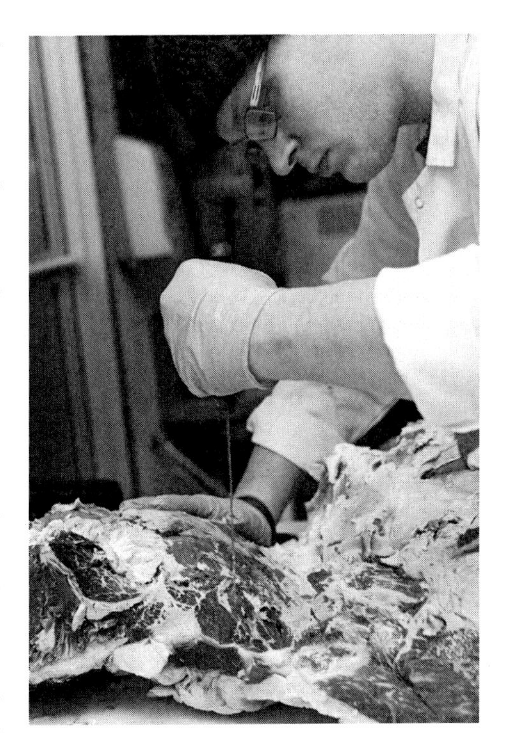

Chop Shop head butcher Tyler Cook cutting meat. *Amy Kalyn Sims.*

One major advantage to getting meat at a butcher shop versus a supermarket meat counter is consumer education, Wright said. "The butcher is always willing and eager to talk about the cuts of meat, what other cuts you might want to substitute and how to select the best cut for the dish you're making."

Wright is committed to supporting local and sustainable agriculture. "Our meat comes from local, family-owned farms," he said. "The animals are pastured and never given hormones or antibiotics. This gives the meat a superior flavor."

In addition to beef, pork and lamb from local farms, Wright also stocks game meats such as bison, elk, venison and quail whenever possible. These are sourced locally if possible, but occasionally the shop has to reach a little farther for some game meats.

Wright says that relationships with local farmers are key to providing local and sustainable meat for his customers. "We're able to tell you everything about the animal—where it lived, what it ate, how long ago it was butchered."

In addition to cuts of meat, the Chop Shop also offers sausages and charcuterie made from the same quality meats as its cuts. "We make our sausages and charcuterie in house in small batches," Wright said. "We carry a variety of fresh sausages and many that are smoked in our on-site smokehouse."

Wright makes good use of the smoker, turning out everything from pulled pork barbecue to chicken wings to meatloaf as daily specials, letting customers know about the day's delicacies via Facebook posts and Tweets. The Chop Shop also offers smoked and cured bacons, pâté, terrines, pastrami and brisket, in addition to deli meats.

WNC Farmers Market

The Western North Carolina Farmers Market is one of four regional farmers' markets in North Carolina, run by the State Department of Agriculture. Housing both wholesale and retail operations, the market is located on Brevard Road and is open 7 days a week, 365 days a year. The retail buildings are open from 8:00 a.m. until 5:00 or 6:00 p.m., depending on the month.

Since its opening in 1977, the market has been a primary source of fruits, vegetables, country hams, local cheeses and stone-ground grits for area chefs and home cooks. The market's three buildings cover thirty-eight thousand square feet and sell to grocery stores, restaurants, chefs and the general public.

Tailgate Markets

Tailgate markets grew from the tradition of farmers bringing their produce into town to sell to city folks who didn't have the space to grow all the fruits and vegetables they needed to get through the winter. Farmers would bring their wagons, and later their pickup trucks, into town and set up on the side of the road. Soon they began to congregate in vacant lots, and the tradition of the tailgate market was born.

Tailgate markets offer chefs, restaurants, foodies and home cooks an opportunity to source local, fresh and often certified organic produce directly from the farmers who grew it.

From one market in the 1980s, Asheville has seen an explosion in tailgate markets, now numbering over a dozen.

This is a listing of tailgate markets located in Asheville and the surrounding area. For information on other markets and to check current locations and hours, visit the Appalachian Sustainable Agriculture Project website at asapconnections.org or fromhere.org.

Asheville City Market
8:00 a.m.–1:00 p.m. Saturdays
161 South Charlotte Street
www.asapconnections.org.

North Asheville Tailgate Market. *Amy Kalyn Sims.*

Asheville City Market South
1:00–5:00 p.m. Wednesdays
2 Town Square Boulevard, Biltmore Park Town Square
www.asapconnections.org

Black Mountain Tailgate Market
9:00 a.m.–noon Saturdays
130 Montreat Road, behind the First Baptist Church in Black Mountain

East Asheville Tailgate Market
3:00–6:00 p.m. Fridays
Groce United Methodist Church's parking lot, at the corner of Beverly and Tunnel Roads

French Broad Food Co-Op Market
2:00–6:00 p.m. Wednesdays
76 Biltmore Avenue, in the parking lot next to the French Broad Food Co-op

Leicester Farmers Market
9:00 a.m.–2:00 p.m. Saturdays
338 Leicester Road, Leicester Landing Shopping Center, behind Zaxby's
www.leicesterfarmersmarket.com

Montford Tailgate Market
2:00–6:00 p.m. Wednesdays
34 Montford Avenue, parking lot of the Asheville Visitors' Center

North Asheville Tailgate Market
8:00 a.m.–noon Saturdays
UNC–Asheville Campus Commuter Lot C. Take Weaver Boulevard to campus entrance at traffic circle; first parking lot on right.

Oakley Farmers Market
3:30–6:30 p.m. Thursdays
607 Fairview Road

Weaverville Tailgate Market
2:30–6:30 p.m. Wednesdays
On the hill overlooking Lake Louise behind the Community Center on Weaverville Highway

West Asheville Tailgate Market
3:30–6:30 p.m. Tuesdays
718 Haywood Road, in the parking area between the Grace Baptist Church and Sun Trust Bank

White Horse Farmers Market
3:00–6:00 p.m. Wednesdays
105 Montreat Road, Black Mountain

Farm Stores and Stands

Many area farms have farm stores or produce stands, ranging from the elaborate ones at Flying Cloud and Hickory Nut Gap Farms to simple affairs that are little more than tables and an awning. Here are a few of the major ones; for a complete list, go to the ASAP Local Food Guide website at www.buyappalachian.org.

Hickory Nut Gap Farm Store
Open: Seven days/week 9:00 a.m.–6:00 p.m. in September and October; Tuesday–Saturday, 10:00 a.m.–5:00 p.m. year round
Where: 57 Sugar Hollow Road, Fairview
Offers: Meat from Hickory Nut Gap; organic fruits and berries in season; jams, jellies and preserves; honey; pickles and preserves; local crafts
Details: www.hickorynutgapfarm.com

Flying Cloud Farm Stand
Open: All day, every day, May–December
Where: 1860 Charlotte Highway, Fairview
Offers: Self-service, honor system farm store offers local organic fruits, vegetables and flowers
Details: www.flyingcloudfarm.net

Looking Glass Creamery
Open: 3:00–7:00 p.m. Thursdays, 11:00 a.m.–5:00 p.m. Saturdays
Where: 57 Noble Road, Fairview
Offers: Goat and cow milk cheeses
Details: www.ashevillecheese.com

Community-Supported Agriculture (CSA)

When a customer connects directly with a farmer to buy an advanced share of the farms crop for the year, it's called community-supported agriculture (CSA).

CSAs provide cash up front for farmers to buy seeds, start plants and generally have cash on hand before harvest time. In exchange, the consumer gets a box each week containing whatever happens to be ready to pick that week. It's a win-win situation, providing the farmer with greater economic stability and the consumer with a season-long supply of local, seasonal ingredients for home-cooked meals.

When ASAP started the Local Food Guide back in 2002, about twelve farms in the guide offered CSAs. In the 2013 guide, the number had increased to ninety-seven, a testament to the popularity of the practice.

For information on local farms that offer CSAs, visit www.buyappalachian.org.

Chapter 7
BEER

The Scots and Scots-Irish settlers who came to the mountains of Western North Carolina in the mid-1700s were far better known for their whiskey than for beer, but the early settlers did brew beer in Asheville.

In the early days, what passed for beer probably wouldn't wash with modern-day Asheville beer drinkers, though. Since the climate in the Blue Ridge wasn't ideal for growing either hops or barley, and cleared land was much better used for corn (which could yield both food and hooch), beer was mostly brewed from locust pods or persimmons. Although most excess corn was made into whiskey (which had more "bang for the buck"), corn was also used by early settlers to make beer.

John Arthur wrote about Mountain Lager, another home-brew of the early settlers, in *Western North Carolina, A History (1730–1913)*: "Mountain Lager Beer or Methiglen, a mildly intoxicating drink, made by pouring water upon honeycomb and allowing it to ferment, was a drink quite common in the days of log rollings, house raisings and big musters. It was a sweet and pleasant beverage and about as intoxicating as beer or wine."

By 1900, Asheville boasted at least eighteen taverns and saloons, most of them located in an area called "Hell's Half-Acre" for its wild activity. Almost all of these places brewed their own house brand of beer and distilled their own whiskey (or bought it legally or otherwise from folks farther up the mountain).

The railroad made transporting beer from the large brewers in the Midwest feasible, and the sudden influx of tourists who came to Asheville would pay

Some of the wide variety of beer available from Asheville's breweries. *Amy Kalyn Sims.*

good money for it. The lighter, fizzy, German-style lagers these breweries produced became popular and edged out the locally produced beer.

All this beer drinking came to a screeching halt in 1907, when Asheville overwhelmingly voted to end all sales of alcohol, largely due to a temperance movement that began late the previous year when an escaped convict named Will Harris decided to mix a quart of whiskey from the Buffalo Saloon with a .303 Savage rifle from Finklestein's Pawn Shop, resulting in six people dead in Hell's Half Acre.

Beer drinking went underground until national Prohibition was repealed by the passage of the Twenty-first Amendment in December 1933.

But boy, did it ever make a comeback starting in the 1990s with the advent of the Asheville craft brew movement.

Beer City, USA

Today, Asheville is home to a dynamic and vibrant beer culture. In addition to the pioneer brewers, the city is home to twelve craft brewers, with more set to open in 2013. This gives Asheville more breweries per capita than anywhere else in the United States.

Home to numerous beer festivals, pub tours and beer tours, at any given time, it's possible to find more than fifty local beers on tap in the city's restaurants and bars, leading Asheville to be named the winner of the Beer City USA poll for four years running.

All this has led to national attention and a reputation as a beer destination. In 2012, national craft brewer Oskar Blues Brewing opened an East Coast expansion in nearby Brevard. The plant produces kegged and canned beer, most notably its Dale's Pale Ale, and is expected to produce fifty-six thousand barrels in 2013. It also offers a free trolley from downtown Asheville to the brewery's tasting room Saturdays and Sundays. Visit oskarblues.com to find out more.

Sierra Nevada will begin test brewing in the summer of 2013 at its new 350,000-barrel production facility in nearby Mills River, with full production slated for the fall of 2013. The facility will employ about 125 people initially.

And in 2015, Colorado-based New Belgium Brewing is slated to open an expanded capacity East Coast facility in the River Arts District, which will increase the company's production to 920,000 barrels per year.

Leading the way to this boom is a community of creative craft brewers.

ASHEVILLE CRAFT BREWERS

Highland Brewing Company

While brewing beer isn't rocket science, it took a nuclear engineer to start the beer revolution in Asheville.

Oscar Wong, who started out in 1994 brewing beer in the basement of Barley's Pizzeria and Tap Room in downtown Asheville with his original brewer John McDermott, is the founder and guiding force behind Highland Brewing Company, Asheville's oldest brewer.

The first three batches he tried went down the drain into the city sewer system, but Wong persevered, and soon Highland Celtic Ale and Highland Galic Ale were taking the city by storm.

The company moved out of the basement and into a real facility in 1993, and in 1997, John Lyda bought out McDermott, becoming Highland's vice-president and head brewer.

In 1998, Highland added a bottling line, and four years later, a second line was added to keep up with demand for the company's products.

In 2005, Highland maxed out production at its facility at 6,500 barrels and moved to its current location in the old Blue Ridge Motion Pictures studio in East Asheville. The new facility increased production capacity to 30,000 barrels per year.

Today, Highland produces both year-round and seasonal beers, and its Gaelic Ale is the most popular beer brewed in North Carolina.

Highland Brewing has a tasting room at its production facility at 12 Old Charlotte Highway, Asheville. Call (828) 299-3360 or visit highlandbrewing.com.

Green Man Brewing

Asheville's second brewery got its start when Jack Eckert, co-owner of Laughing Seed vegetarian restaurant with then-wife Joan Clincy-Eckert, decided the unused basement would be a good spot for a pub. Thus, in 1997, Jack of the Wood was born on Patton Avenue downtown.

Named for a synonym for the Green Man, a Celtic wood spirit, Jack of the Wood was only a pub at first until a baker who was renting kitchen space mentioned that her boyfriend was a brewer.

Soon Joseph Rembert, the brewer boyfriend, was hard at work churning out Green Man Gold, a blonde ale Rembert made in converted dairy tanks and pumped directly to the taps. A double batch made fresh every day was just enough to supply the pub and leave a keg or two for the Laughing Seed.

Rembert left Green Man in 2001 to start French Broad Brewing Company and was replaced by Mike Duffy, who came from Boston's Harpoon Brewing. Several brewers came and went until John Stuart, Green Man's current brewer, took over.

The brewery outgrew its space at Jack of the Wood and, in 2005, moved to a larger space on Buxton Avenue. It opened a taproom at the new brewery, and locals promptly dubbed it "Dirty Jack's." The name stuck.

Jack and Joan sold Green Man in 2010 to Dennis Thies, a Florida native who moved to Asheville after his family sold their beer distributorship. In 2012, Dennis bought the building next to Dirty Jack's and added a thirty-barrel production facility, tripling Green Man's production.

The Green Man Brewing tasting room known as Dirty Jack's is at 23 Buxton Avenue in Asheville. Call (828) 252-5502 or visit greenmanbrewery.com. Or sample the beers at Jack of the Wood, 95 Patton Avenue, Asheville.

Asheville Brewing Company

The third of the city's brewing pioneers is Asheville Brewing Company.

In 1998, a company out of Portland, Oregon, opened Two Moons Brew 'n' View on Merrimon Avenue with brewer Doug Riley handling the brewing operation. Despite a good combination of beer, handmade pizza and cheap movies, the business was tanking.

Enter Mike Rangel, a restaurant consultant hired by the Brew 'n' View's owners to sell the joint or shut it down. Rangel ended up buying the business and is now its president and co-owner.

Rangel, along with Leigh, his wife at the time, went in with Riley and mortgaged their homes to get enough money to buy the failing business. They set about fixing the food and service problems that plagued the Brew 'n' View. After a name change to Asheville Pizza and Brewing, a massive cleanup and a little remodeling, the place was ready to open.

Some of Asheville Brewing Company's early beers were Roland's ESB (now called Ninja Porter), Shiva IPA, J.T.'s Oatmeal Stout and Pisgah Pale.

In 2006, the company opened a second location on Coxe Avenue that offers beer and pizza but no movies. It also contains an additional production facility that significantly increases Asheville Brewing Company's production. After this facility opened, ABC started selling kegs to other restaurants and bars in the region and, in 2011, opened a canning line, becoming the first brewer in Asheville to can beer.

Visit Asheville Pizza and Brewing at 95 Merrimon Avenue. Asheville Brewing is at 77 Coxe Avenue. For more information, visit ashevillebrewing.com.

French Broad Brewing Company

French Broad Brewing Company, named for the river that flows through Asheville, is the last of the city's pioneer breweries.

Founded in 2001 by former Green Man brewers Jonas Rembert and Andy Dahn, French Broad Brewing Company began turning out beer in a former warehouse on Fairview Road.

The brewery offers a taproom with a small stage in front of the tanks and live music several nights a week, something unique to French Broad.

Soon after opening the brewery, Andy left to pursue other interests but returned in 2006 as the company's general manager. He also owns a stake in the cooperative group that owns French Broad Brewing. Jonas left shortly after Andy returned and now lives outside the United States; brewing has been taken over by Chris Richards, Aaron Wilson and John Silver.

French Broad's Goldenrod Pilsner was one of the brewery's earliest and most popular brews and still leads sales. Its Gateway Kolsh is also popular, along with 13 Rebels ESB and Rye Hopper Pale.

French Broad Brewing (and its tasting room) is at 101 Fairview Road, Asheville. Call (828) 277-0222 or visit frenchbroadbrewery.com.

Wedge Brewing Company

Located in Asheville's River Arts District (RAD), this combination brewery, tasting room and sculpture garden is a favorite weekend hangout.

Owner Tim Schaller and brewer Carl Melissa opened Wedge in 2008 and have been going strong ever since. Named after the Wedge Building, the brewery's home, Wedge Brewing produced 1,300 barrels in 2012, almost all of that output being served at the taproom.

Wedge's beers range from Gollum, a Belgian style, to Iron Rail, a hoppy IPA. It also offers various seasonal beers as well.

Wedge Brewing is at 125B Roberts Street, Asheville. Call (828) 505-2792 or visit wedgebrewing.com.

Oyster House Brewing

In 2008, Billy Klingel was a bartender at the Lobster Trap in downtown Asheville when he first heard of a beer called oyster stout. An oyster freak and dedicated home-brewer, Klingel went through sixty-two batches of oyster stout prototypes before coming up with his version of the brew.

Klingel invited Lobster Trap owner Amy Beard over to sample the new brew and then convinced her to offer it at the restaurant. After a nine-month permitting process made more challenging because he used a portable brewing system, the beer, dubbed Moonstone Oyster Stout, began flowing at the Lobster Trap in the spring of 2009. It was joined by Moonstone IPA, Patton Ave. Pale, Dirty Blonde and Upside Down Brown. And in case you're wondering, the name isn't just a gimmick: each fourteen gallons of beer contains five pounds of oysters.

In the summer of 2013, Oyster House moved out of the Lobster Trap and into its own space at 625 Haywood Road in West Asheville. The new spot is a tiny brewpub offering food and oyster-flavored beers. Call (828) 575-9370 or visit oysterhousebeers.com.

Lexington Avenue Brewing Company

Mike Healy and Steve Wilmans teamed up to open Lexington Aveneue Brewing (LAB) in 2010 as a gastropub, a brewpub featuring gourmet food and craft-brewed beer.

In 2006, Wilmans, a recording engineer from Seattle, had just sold his studio there and was in town helping Healy move to Asheville. He saw an old church for sale, realized it would make a righteous recording space and made the move himself, opening Echo Mountain Studio later that year.

A running joke that the pair ought to own a bar led to the purchase of the old T.S. Morrison store at 29 Lexington, and after a year and a half of renovations, the pair opened LAB in 2010.

They hired Ben Pierson, formerly of Green Man, to handle the brewing chores and Jason Roy, a former executive chef at the Greenbriar Resort in Colorado, to run the kitchen. Soon they had customers spilling out the doors. LAB became the first Asheville brewery to win an award for its beer, taking a bronze medal at the 2011 Great American Beer Festival.

LAB is at 39 North Lexington Avenue, Asheville. Call (828) 252-0212 or visit lexavebrew.com.

Thirsty Monk Brewing

Barry Bialik wins the prize for "Most Distance Traveled to Open an Asheville Brewery," hands down. The former real estate developer and alternate newspaper publisher heard good things about Asheville all the way up in Anchorage, Alaska, and decided to check things out. In 2007, he flew down for a visit, liked what he saw and put an order in on a house.

A fondness for Belgian beers led Bialik to open the Thirsty Monk, a bar specializing in Belgian brews. Finding the basement of a building on Patton Avenue for rent that formerly housed a hookah bar, Barry instead bought the entire building and opened the Monk in 2008 in the basement. A few months later, his upstairs tenant went out of business, so he opened the upstairs as an American craft brew bar, making the Monk one of downtown's most popular watering holes. Bialik tries to keep the Monk's forty taps flowing with as many rare and unusual beers as possible.

In 2010, Bialik and regular Monk customer (and home-brewer) Norm Penn came up with a scheme to put a one-barrel brewing system into the Monk's second location in South Asheville's Gerber Village, and the Monk's first brew debuted in October 2012. The Monk releases one or two beers a week, usually on Thursdays.

In the summer of 2013, the Monk was slated to open a third location, a much bigger brewing facility and pub at Biltmore Park.

Thirsty Monk downtown is at 92 Patton Avenue, Asheville. A second location is at 20 Gala Drive, Asheville. Look for the third location in Biltmore Park, South Asheville. Visit monkpub.com.

Altamont Brewing

Opened as a taproom and music venue on West Asheville's trendy Haywood Road in 2011, Altamont Brewing Company started brewing its own beers in December 2012.

Brewer Gordon Kear has turned out an assortment of ales, including IPAs, American pale ale, a gold, a red and a rye.

Altamont Brewing is at 1042 Haywood Road, Asheville. Call (828) 575-2400 or visit altamontbrewingcompany.com.

Wicked Weed Brewing

Located on Biltmore Avenue next to the Orange Peel Music Hall, Wicked Weed refers to hops, not that other…you know…

With a restaurant on top and tasting room on bottom, the Weed quickly made a name for itself after opening in December 2012 for gastropub fare and good beers, including Belgian sours and barrel-aged ales. Many feature irreverent names such as Zealot IPA, Infidel Porter and Tyrant Double Red. Be aware, there are often lines out the door to get in.

Wicked Weed Brewing's restaurant and tasting room are at 91 Biltmore Avenue, Asheville. Call (828) 575-9599 or visit wickedweedbrewing.com.

Hi-Wire Brewing

Located in the old Craggie Brewing location (which closed in 2012) in what has become Asheville's brewing district, Hi-Wire opened in June 2013 and features four year-round beers (a pilsner/lager, a brown ale, an IPA and a pale ale) plus a selection of changing seasonal ales.

Hi-Wire Brewing is at 197 Hilliard Avenue, Asheville. Call (828) 575-9675 or visit hiwirebrewing.com.

The Future

As of the early summer of 2013, Asheville is set to welcome at least three more craft brewers and a heavy hitter to town.

Burial Beer will open at 40 Collier Street with a one-barrel nano system. Its owners plan to build a much bigger "farmhouse" brewery in the Asheville area. Twin Leaf Brewing will have a ten-barrel system on Coxe Avenue, and Catawba Valley Brewing of Morganton is building an expansion Asheville brewery at 2 Fairview Road at the old Rankin Patterson Oil site near French Broad Brewing. The Asheville location will include a tasting room, patio and seven-barrel brewhouse.

Chapter 8

WINE AND 'SHINE

Wine has always played a part in Asheville food history, although until modern times a tiny one. The earliest settlers made wine from the native wild grapes and blackberries they found, especially the Waldensians, an Italian religious sect that settled at the foot of the Blue Ridge in Valdese, North Carolina. But grape production was not a part of Asheville's early history.

Then George Vanderbilt came to Asheville to live at Biltmore House. As he brought along a classically trained and discriminating palate for fine wine, soon his staff was hard at work to fill the estate's wine cellar with appropriate vintages.

Wine purchases for the estate were handled by Thomas Morch, Mr. Vanderbilt's secretary in New York. He ordered much of the wine for the estate from France, especially sherry, Madeira, Chablis, Volnay, Beaune, Hock, Claret, Bordeaux and champagne, ordering frequently from Alexander Morten, president of Morten & Co. Wine Merchants and Importers in New York City.

The Biltmore Winery

In 1971, William A.V. Cecil, George Vanderbilt's grandson, planted a small, experimental vineyard on Biltmore Estate property as part of his efforts to make the estate self-sustaining and more profitable. The first planting, thirty

Jamie Ager is a fourth-generation farmer and manager of Hickory Nut Gap Farm. *Amy Kalyn Sims.*

Walter Harrill of Imladris Farm gave up test tubes for jelly jars and a successful career supplying jams to Asheville restaurants. *Amy Kalyn Sims.*

John and Julie Stehling, founders of Early Girl Eatery, were early advocates of sourcing from local farmers. *Amy Kalyn Sims.*

Cinnamon rolls, West End Bakery. *Amy Kalyn Sims.*

Above: Biltmore Estate Winemaster Bernard Delille and Winemaker Sharon Fenchak in the barrel room. *Photo courtesy of the Biltmore Company.*

Right: Chef Damien Cavicchi of Biltmore Estate making pasta. *Amy Kalyn Sims.*

Roasted salmon with gremolata, fresh kale salad, sesame ginger tofu and shepherd's salad from Laurey's Catering. *Amy Kalyn Sims.*

Organic radishes at West Asheville Tailgate Market. *Amy Kalyn Sims.*

Roasted brussels sprouts with shaved parmesan, Laurey's Catering. *Amy Kalyn Sims.*

Buttermilk fried chicken with crawfish mac and cheese, toasted bread crumbs, chicken giblet and egg gravy and local pea shoots from Seven Sows Bourbon & Larder. *Amy Kalyn Sims.*

Above: Potato gnocchi with grilled asparagus, guanciale and poached egg. *Amy Kalyn Sims.*

Left: Esqueixada de Montaña with Sunburst Farms trout served raw with fresh tomato, black olive, sweet onion and a lemon-tarragon vinaigrette from Cúrate. *Amy Kalyn Sims.*

Chef William Dissen seasons snapper at the Market Place. *Amy Kalyn Sims.*

Chef William Dissen plates a dish at the Market Place. *Amy Kalyn Sims.*

Pan-roasted wild snapper, roasted spring vegetables, watercress, pickled radish and chervil salad from the Market Place. *Amy Kalyn Sims.*

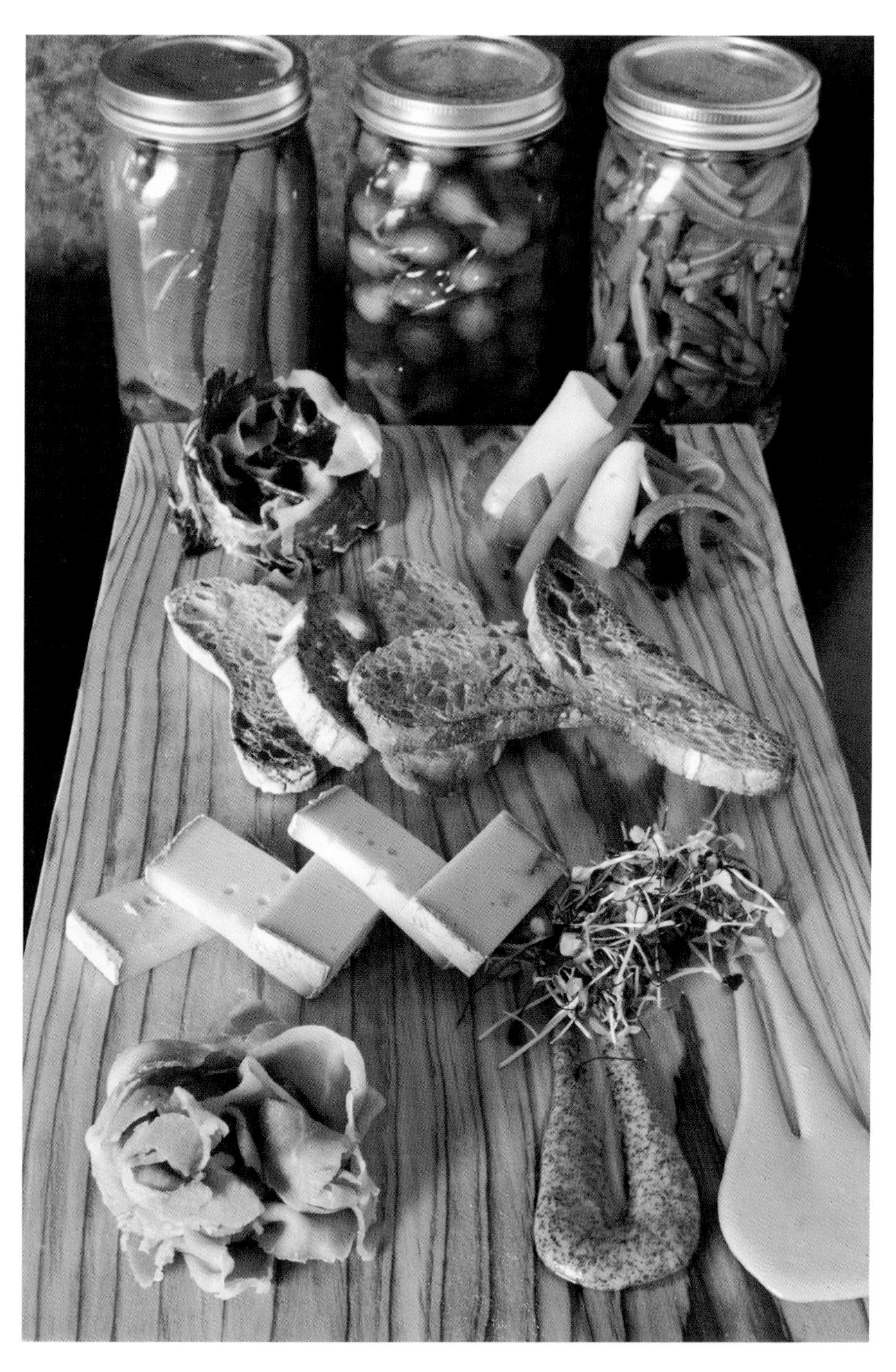

Assortment of shaved country hams with pickled vegetables, toast, black mustard, Texas Pete mayonnaise and cheese from Seven Sows Bourbon & Larder. *Amy Kalyn Sims.*

Not Your Mama's Meatloaf with local hormone-free, grass-fed beef; bacon; and rosemary tomato shallot gravy served with scratch-made mac'n' cheese and fresh asparagus from Tupelo Honey Café. *Amy Kalyn Sims.*

Ultimate Veg: organic lemon tahini chickpea hummus with roasted red peppers, swiss cheese, lettuce, tomato, cucumber, sprouts, red onion and house-made hot sauce from West End Bakery. *Amy Kalyn Sims.*

Right: Calico Bay scallops on creamy wild mushroom rice from Chestnut. *Amy Kalyn Sims.*

Below: Imladris Farm raspberry jam. *Amy Kalyn Sims.*

Tyler Cook, head butcher at Chop Shop Butchery, with a tray of steaks. *Amy Kalyn Sims.*

Petit fours, A-B Tech Pastry Arts class. *Amy Kalyn Sims.*

Vineyards at Biltmore Estate. *Photo courtesy of the Biltmore Company.*

acres of muscadine grapes on land located below the house, was somewhat lacking, in Mr. Cecil's opinion.

He consulted with experts at North Carolina State University and Cornell University and tried growing French-American hybrid grapes. He was again not satisfied with the quality of his efforts.

After viniculture experts at the University of California–Davis told Cecil that growing grapes in Western North Carolina was impossible, he ramped up his efforts and went to France. The French experts told Cecil that viniculture was possible on the estate but would take time and money.

While in France, Cecil hired Philippe Jourdain as the estate's first winemaker. Jourdain, a sixth-generation winemaster from Provence, had operated a family vineyard. He was also a respected teacher of viticulture and oenology, having taught at the Lycee Agricole in Carcassonne.

In 1978, the first vinifera grapes were planted on the west side of the estate, and the following year Jourdain oversaw the first wine offered to the public under the Biltmore label.

By the early 1980s, Biltmore had switched to *vitis vinifera* grapes, also planted on the estate's west side. The Biltmore Estate Wine Company was established in 1983. Soon afterward, construction began to turn the Richard Morris Hunt–designed dairy barn into a production winery. The early years also brought the realization that cold winters and hot summers would occasionally make for lean harvests of estate grapes, and alternate sources in California were established to supplement the estate's production when necessary.

By 1985, the winery had opened to the public, and wine tastings began, making the winery a popular spot on the estate. Today, it's the most visited winery in the United States, with more than 600,000 visitors each year.

Bernard Delille joined the winery as winemaker in 1986 and was named winemaster after Jourdain retired in 1996. The following year, Biltmore wines were served at the White House as part of a state dinner. In 1999, winemaker Sharon Fenchack joined Delille. By 2001, the estate was producing in excess of one million bottles annually, and in 2010 the winery celebrated its twenty-fifth anniversary.

Today, the vineyard is located on the west side of the French Broad River, which flows through the estate. It consists of ninety-four acres of grapes.

The Winery at Biltmore Estate is the most visited winery in the United States. *Photo courtesy of the Biltmore Company.*

Grapes grown on the estate today include cabernet sauvignon, cabernet franc and merlot for red varietals and Riesling, chardonnay and viognier for white varietals. Other varietals and additional grapes are sourced from some of the nation's finest vineyards in North Carolina, California and Washington State.

The winery production facility has grown to ninety thousand square feet and produces 250,000 cases of wine each year, placing Biltmore among the top wine producers in North Carolina. At the winery, grapes are weighed, de-stemmed and "crushed" at the crush dock. Wines are fermented in one of seventy-five tanks and aged in oak barrels or steel tanks, bottled in the bottling room and sampled in the tasting room, which was once the dairy milking parlor.

Find more information at biltmore.com/our_wine.

ADDISON FARMS VINEYARD

For decades, Biltmore held the title as the only vineyard in Buncombe County. But a newcomer has brought winemaking to the rolling farmland of Leicester, a rural Buncombe County community west of Asheville.

Addison Farm Vineyard is a family farm that has been in Jeff Frisbee's family for more than seventy-five years. Addison Farmer and his parents purchased the property in 1937; the Farmers purchased an adjacent property in 1955.

In 2009, Jeff and Diane Frisbee decided to convert what had been a tobacco and cattle farm into a vineyard and winery, planting the first acre of a planned ten-acre vineyard. Subsequent plantings in 2010, 2011 and 2012 have given the Leicester farm six of the planned ten acres of grapevines. Current varieties of grapes include cabernet sauvignon, cabernet franc, sangiovese, Montepulciano, petit verdot and petit manseng. For 2013, they are adding an additional half acre of petit manseng grapes.

Frisbee and his family produced their first crush in 2010 from fruit purchased from other regional growers. In 2011, they crushed the first grapes from their own vineyard in addition to grapes from other regional vineyards.

Currently, Addison Farm offers six varieties, including Gwinn, a white made from traminette and chardonnay grapes; Gratitude, a red dessert-style wine made from chambourcin grapes; Front Porch Red 2010 made from sangiovese grapes; Orion, a white featuring chenin blanc and chardonnay grapes; Coming Home 2011, a cabernet sauvignon; and Front Porch Red NV, made from chambourcin and sangiovese grapes.

Future releases still in the barrel include 2010 sangiovese; 2011 chambourcin; 2012 chardonnay; 2012 Montepulciano; 2011 cabernet sauvignon; 2012 cabernet sauvignon; and 2012 Bordeaux blend consisting of cabernet sauvignon, cabernet franc and petit verdot.

Addison Farm opened a tasting room in late 2012, large enough for events as well as tastings and retail sales. In 2013, it purchased a 350-gallon stainless steel tank to increase its white wine production. It plans to increase its crush from 1,000 gallons in 2012 to 6,000 gallons by 2015.

"Building the vineyard and winery gives our family the opportunity to continue the farming tradition on this piece of property through at least the fifth generation of our family," Jeff said. "The family farm is a key component of our community in Western North Carolina. Too many family farms cease to exist after a generational change as the property falls prey to residential and commercial development. We do not want to see that happen here."

Addison Farm Vineyard is at 4003 New Leicester Highway, Leicester, North Carolina. Call (828) 581-9463 or visit addisonfarms.net.

MOONSHINE

There are few things in food history that can be traced to an exact date. Moonshine is one of them.

Moonshine was created by order of no less than George Washington, who, in order to pay off France for all the money the United States borrowed to fight the Revolutionary War, placed a federal excise tax on all spirits distilled in the United States after March 3, 1791.

Prior to this date, there was no such thing as moonshine, only corn whiskey. After March 3, anyone who didn't pay tax on their liquor (and that included most people in Western North Carolina) was making illegal liquor, or moonshine.

For some, it was the amount of tax levied by the government that kept their whiskey making on the other side of the law. As one mountaineer explained to Horace Kephart, author of *Our Southern Highlanders* (1913): "Nobody refuses to pay his taxes for taxes is fair and squar'. Taxes costs mebbe three cents on the dollar; and that's all right. But revenue costs a dollar and ten cents on twenty cents' worth o' liquor; and that's robbin' the people with a gun to their faces. Now, yan's my field o' corn. I gather the corn, and shuck hit, and grind hit my ownself, and the woman she bakes us a pone o' bread to eat—and I don't pay no tax, do I? Then why can't I make some o' my corn into pure whiskey to drink, without payin' tax?"

Corn liquor was a valuable trading commodity in a region where cash money was scarce and rough mountain roads made it hard to transport loads of corn to market. Someone who had corn liquor could trade for powder and shot, flour, gingham and calico for the lady of the house, chewing tobacco or whatever else he needed or wanted to make life in the wilderness more pleasant.

Oftentimes would-be moonshiners went in with family or friends to pool resources to purchase a used still or to procure the parts to assemble a new one. The basic still consisted of a sealed copper kettle or pot in which to cook the mash and a copper condensing coil called a worm. The distillation process for moonshine required a fire to "cook" off the alcohol. Since this produced large amounts of smoke and steam, the process was carried out at night far back in the woods, where the fire was indistinguishable from any normal campfire. Since the operation was carried out by the light of the moon, the word "moonshine" entered the American vernacular as both noun and verb.

Since Asheville was "the big city," moonshining was limited to areas outside the city and up in the hills—until recently.

Troy & Sons Distillery

Close your eyes and get a mental image of Asheville's most successful producer of corn liquor hard at work making white lightning. Now, let's bust it wide open.

Troy Ball's still isn't cobbled together from a wash tub, some copper pipe from the Feed and Seed and a radiator from a '64 Ford Fairlane. No, sir, it was designed by an engineer and made in Germany.

Troy also has a license from the United States government saying it's OK to make 'shine, as long as they get a cut off it. And finally, if you're picturing some old coot with a long white beard, well, Troy shoots a hole in that stereotype too. You see, he's...well, he's a she.

Troy Ball is the CEO of Asheville Distilling Company, producers of Troy & Sons whiskey. Along with her husband and business partner, Charlie, Ball has brought moonshining out in the open and made Asheville whiskey the darling of Southern society.

Troy Ball, CEO of Asheville Distilling Company and the Troy in Troy & Sons. *Photo courtesy of Troy & Sons.*

Like many others who "cook up" a good recipe, Troy credits much of her liquor's success to a seventh-generation local farmer, John McEntire, who grows corn on a family farm just south of Old Fort. The corn, called Crooked Creek, is an American white variety first planted by McEntire's family in the mid-1800s. Ball uses it exclusively and credits the corn with making the whiskey taste mellow and flavorful.

The idea for Troy & Sons began in 2004, when several Madison County neighbors welcomed the Balls, recently relocated from Texas, with under-the-counter gifts of pint Mason jars filled with the local 'shine. After a few sips, the Balls realized that there might just be a market for a refined, potent (and legal) version of good old mountain moonshine.

So with a sixty-gallon still located in John McEntire's red barn, Troy and Charlie set out to learn how to make quality corn liquor, experimenting with and modifying recipes from the North Carolina State Archives. They found a fan and mentor in Oscar Wong of Asheville's Highland Brewing, who praises Troy & Sons 'shine for its mildness.

"Other moonshine I've tried always had a harshness to it that you just came to expect," Wong said. "This is smooth and sippable by itself."

After making many batches of white liquor, Troy and Charlie discovered that the beginning of the run, what old-timers call the "head," contained all kinds of chemical nasties such as acetone, methanol and aldehydes that not only aren't very good for you, but they give moonshine its "Great Balls of Fire" burn. They eliminated these nasties from the spirits, and the result is their Platinum Moonshine, a clear, clean product with hints of vanilla and an uncharacteristically mild flavor.

They also take some of the 'shine and age it in bourbon barrels to produce Troy & Sons Oak Reserve, an aged moonshine with a slight amber color and a smooth finish.

In the spring of 2011, Troy and Charlie moved into a new production facility in the old Southern Railway wheelhouse next to Highland Brewing Company. The facility features a five-thousand-liter still Charlie designed and had manufactured to Troy & Sons specifications in Germany. The couple runs six distillations a week and offers samples of their handiwork at a gleaming copper bar in their new tasting room.

Their whiskies are available in package stores and restaurants in six states and have been featured on the *CBS Morning News* and MSNBC, as well as in *Bon Appetit*, *Southern Living* and *Garden & Gun* magazines.

Troy & Sons Distillery is at 12 Old Charlotte Highway, Asheville. Tasting tours are offered at 5:00 and 6:00 p.m. Thursday–Saturday. Call (828) 575-2000 or visit Troyandsons.com.

Chapter 9
CULINARY EDUCATION

Ask any local chef to name one distinct advantage Asheville has over most cities its size when it comes to attracting restaurants, and chances are Asheville-Buncombe Technical Community College will be the answer.

Mercifully shortened from its full name to A-B Tech, it has been successfully turning out chefs and bakers for area and national restaurants since the late 1960s, and area chefs credit this pool of fully trained and kitchen-ready talent with helping Asheville maintain its culinary edge.

The A-B Tech culinary program was the vision and passion of Robert G. (Bob) Werth, a French-born and classically trained chef who started trying to convince people that the local community college needed to turn out not just nurses and auto mechanics but also chefs trained in the classic French style of cooking.

Given the dearth of Western North Carolina restaurants in the late 1960s to employ these newly minted chefs, he must have seemed mad as a March hare, but as anyone who ever talked with him about cooking and food can attest, Bob was as charming as he was persistent.

In 1968, A-B Tech enrolled its first four students and soon graduated two of them. The program grew exponentially as word spread of the quality of the education available at the tiny and inexpensive school, and soon students came from all over the United States to take advantage of the program.

"You wonder how Asheville became part of the food scene, and it comes back to the development of Asheville in terms of the world of hospitality and Chef Werth," said Sheila Tillman, associate dean of A-B Tech's Hospitality

The culinary program at A-B Tech is a fertile training ground for talent for local and national restaurants. *Amy Kalyn Sims.*

Education Department. A former student of Werth, Tillman took over the program when he retired in 1994. "He was dedicated to the students way before it was our mission statement," she said. "He loved having students come back and see him."

Students in the culinary program are trained in the classic French brigade system, with the kitchen under the command of second-year students who help manage and train the first years, thus developing the skills the second years will need when they are on their own, running the kitchen at a restaurant or resort.

Students are immersed in the program from the first day, always wearing chef coats and tall hats when in class. Their studies begin with sanitation and kitchen safety, basic culinary and knife skills. Since a chef must also be able to keep food costs down, practical math and basic computer skills are also taught early on in the program, along with a class on food and beverage cost control.

The future chefs learn everything from charcuterie (making sausages and terrines) to baking, cake decorating and ice sculpture, all under the watchful eye of experienced instructors and culinary educators who have worked at some of the country's top restaurants and resorts.

The summer between the first and second year is devoted to full-time work at a restaurant or resort, where the students are exposed to the hard work and long hours that make up the life of an executive chef. At the end of the summer, they have a better idea of what lies ahead at the completion of the program.

After another year covering everything from advanced culinary skills to human resources management, graduates are presented with a two-year associate's degree and offered assistance finding a position at a suitable restaurant, resort or hotel.

Some of the places recent graduates have found employment include the Inn on Blackberry Farm, Cherokee Town and Country Club in Atlanta, various Ritz-Carlton properties and scores of local restaurants.

Today, the program enrolls 60 to 70 students per year and has a total enrollment of about 160 students at any one time. The students regularly compete with other culinary programs across the South and almost always bring home top honors. They hone their skills in five state-of-the-art kitchens, including an elevated lecture hall demonstration kitchen.

"There's no doubt that A-B Tech's culinary program is a great asset to chefs who plan on opening restaurants here," Tillman said. "They know that finding a good source of trained, skilled labor is one problem they don't have to worry about with our graduates."

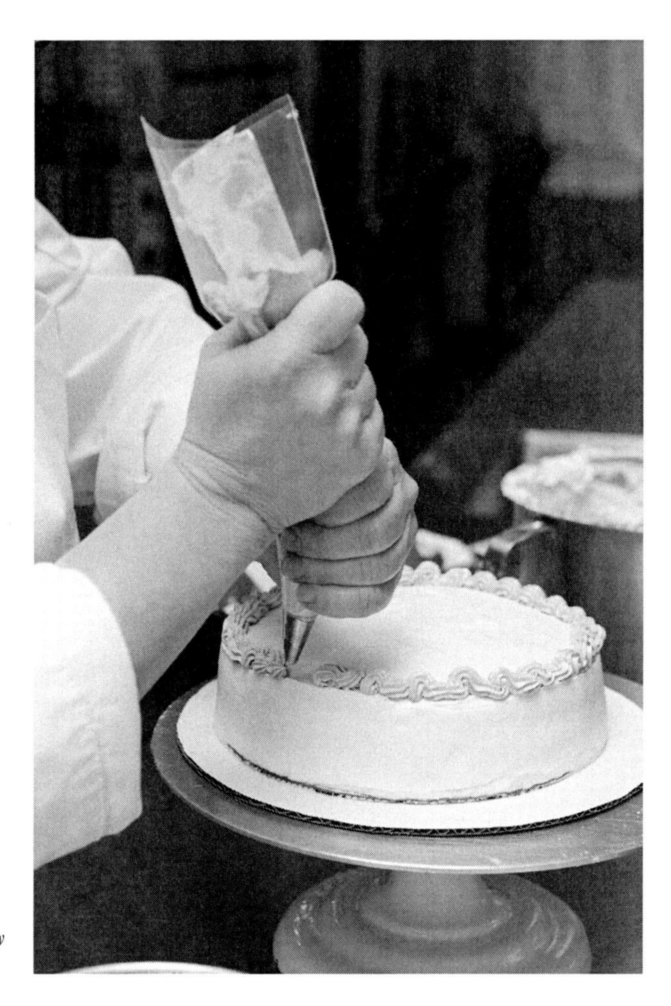

A-B Tech students learn pastry skills as part of a well-rounded culinary education. *Amy Kalyn Sims.*

In the spring of 2013, A-B Tech advertised for a new faculty position for someone to oversee, teach and develop programs for curriculum and non-curriculum students in brewing, fermentation and distillation, a nod to Asheville's role as a leader in brewing and distilling.

What's more, you can dine on student-made meals as they hone their skills. Lunches and dinners are staged by students most Thursdays in the fall and spring semesters in the school's Magnolia and Fernihurst building dining rooms.

ASHEVILLE HIGH SCHOOL CULINARY PROGRAM

Asheville has only one high school to serve students who live in the city limits, and it is an imposing campus, larger than some small colleges. For students who want to learn culinary skills for an entry-level job in the city's hospitality industry or just how to cook something other than mac and cheese from a box, the school offers several culinary classes.

From the basic foods class, which covers nutrition and food safety, to a vocational concentration in culinary arts that gives a more comprehensive introduction to the field, students at Asheville High can get the basics for a culinary career or a lifetime of good home cooking.

GO KITCHEN READY

Green Opportunities is a nonprofit that works with low-income adults who face barriers to employment. The Kitchen Ready program, under the management of veteran chef and restaurant owner Mark Rosenstein, is a culinary training program that prepares students for employment in the restaurant business. Kitchen Ready began as a pilot program funded by the Asheville Independent Restaurant Association (AIR), and after its initial success, the initiative was added as a permanent part of Green Opportunity's program.

Classes cover basic food safety, sanitation, basic culinary skills, baking, food vocabulary and kitchen math. Students also learn life skills such as interviewing techniques, résumé writing, job search techniques, job retention skills and personal finance management. The students also learn the use of local food products and fresh produce.

The program, which began as a joint venture between Green Opportunities; Goodwill Industries of Northwest North Carolina, Inc.; Asheville City Schools Foundation; A-B Tech; the Asheville Independent Restaurant Association; and MANNA Food Bank, enrolled its first class in March 2012 on the campus of William Randolph School. Students receive 192 hours of hands-on kitchen instruction and 96 hours of classroom instruction over a twelve-week period. The training produces ready-to-eat meals for distribution through MANNA Food Bank and before- or after-school meals at William Randolph School.

Blue Ridge Food Ventures

While not entirely an educational program, Blue Ridge Food Ventures provides a valuable service to Asheville's culinary community.

Launched in 2005 as the brainchild of New Orleans native MaryLou Sturgi, Blue Ridge Food Ventures offers prospective culinary entrepreneurs interested in the specialty foods industry classes, incubation help, product development help and an eleven-thousand-square-foot test and production kitchen.

Using its facilities and equipment, both of which can be rented by the hour, local farmers, chefs and food entrepreneurs can develop new food-based products and business ventures without the expense of equipping a production kitchen. Imladris Farm uses its production kitchen to make its renowned blackberry, blueberry and raspberry jams, as do other Asheville food businesses, such as Lusty Monk Mustard.

Chapter 10
THE YOUNG GUNS

Asheville's culinary destiny lies largely in the hands of a new generation of chefs who grew up with twenty-four-hour Food Network on their TVs, who are seldom without Smartphones in their pockets and who Tweet and Facebook nearly all their culinary moments. Most of them weren't even born in 1979 when Mark Rosenstein opened the Market Place.

But for what they lack in age and tradition, they make up in passion, dedication and raw talent. Spend a few minutes with any of them talking food, and you'll find that the next generation is going to set the culinary scene on fire.

Katie Button
Executive Chef/Co-Owner
Cúrate

No one embodies the newfound national respect Asheville's food scene has garnered more than Katie Button, executive chef and co-owner of Cúrate.

In the span of a few months in 2013, Button was nominated as *Food & Wine* magazine's Best New Chef in the Southeastern region and named a semifinalist for the prestigious James Beard Foundation's Rising Star Chef of the Year award.

Although Katie grew up around food (her mother was the chef/owner of a catering company), Button was headed for a far different career.

"I graduated from Cornell University with a degree in chemical and biomolecular engineering and went on to complete a master's degree at L'École Centrale in Paris," Katie said. "I didn't realize that I needed a career change until I was about to start my PhD."

Chef Katie Button is a two-time James Beard Foundation Rising Star Chef nominee. *Amy Kalyn Sims.*

Button took a job as a server at Café Atlantico/Minibar in Washington, D.C., where she "fell in love with restaurants, food, cooking and Felix Meana," Katie said.

Meana, a chef de rang (service manager) from El Bulli, Ferran Adria's famous restaurant in Spain, introduced Button to the flavors of Spain and to the owner of El Bulli.

Katie spent three months as a server at El Bulli, deciding in the process that she wanted to be in the kitchen creating rather than serving. She approached El Bulli's pastry chefs and asked, if she got a year's worth of kitchen experience under her belt, could she return and train with them. They agreed, and after some time working with Johnny Iuzzini at Jean Georges in New York and then at the Bazaar by Jose Andres in Los Angeles, she returned for an apprenticeship at one of the finest restaurants in the world.

In March 2011, Button and her family (Meana manages the front house operations and does double duty as Katie's husband, while her mother, Liz, is general manager and dad Ted handles the financial end) opened Cúrate, choosing Asheville for its size and because "it felt like home."

Buzz about the new restaurant, which featured Spanish tapas and wines, was almost instantaneous, starting on Chowhound, Facebook and local blogs and spreading to the local media. Soon magazines in Charleston and Charlotte were singing Button's praises, followed by *Southern Living*, *GQ* and *Food & Wine*. In November 2012, Button was chosen to cook a Spanish holiday meal at New York's James Beard House. She was nominated for the Beard Foundation's Rising Star Chef award in 2012 and 2013.

Button credits Asheville itself with driving the dynamic food scene that has sprung up in the past few years. "It's the community of Asheville that inspires the food scene. The diversity of people, art, music, culture. The community demands the support of local farmers and the local food scene and all small businesses," she said.

"It is amazing to look at the number of independently owned businesses and restaurants. Many people, like us, seem to pick Asheville as a place that they want to live and follow their passion. They make their own dreams come true," she said. "That is what makes Asheville unique."

Katie credits area farmers with helping Cúrate keep the finest local foods on its tables. "We are developing great relationships with our farmers. I have had a few farms adjust their production of certain ingredients to meet our needs," she said. "I've also had farms plant new ingredients just for us. That kind of personalization is amazing."

Adjusting a menu to fit the growing season can be a challenge, but Katie has found ways to work around the challenges.

"I try to be flexible and supportive of the farms I work with. It is definitely a give and take," she said. "I have a farm that grows us a mix of beautiful petite greens. When they came to me and said they were going to halt production through the winter, I created a salad that didn't rely on their greens, using other ingredients instead. But when they figured out a way to make the greens work throughout the winter, I changed my menu plan to support them through the winter."

Cúrate is at 11 Biltmore Avenue, Asheville. Call (828) 239-2946 or visit www.curatetapasbar.com.

Elliott Moss
Executive Chef/Co-Owner
Ben's Tune Up
Buxton Hill

If you didn't know Elliott Moss and saw him on his day off wearing his timeworn jeans, faded T-shirt under a flannel shirt and ever-present ball cap, it probably wouldn't surprise you to learn he's one of the guys who owns Ben's Tune Up.

Elliott Moss is a self-taught chef who garnered critical acclaim and a James Beard nomination while executive chef at the Admiral. *Amy Kalyn Sims.*

The real surprise would come when you learned that Ben's Tune Up is one of the most eagerly anticipated new restaurants in town, and Moss isn't a mechanic—he's a 2013 James Beard Award semi-finalist for Best Chef in the Southeast.

Moss grew up in Florence, South Carolina, a town noted more for its popularity as a pit stop for people en route to Myrtle Beach than as a dining destination. But in the South Carolina Lowcountry, Moss developed an appreciation for food that would serve him well in the years to come.

Moss spent his boyhood days on his grandparents' farm, shucking corn, shelling beans and playing with the pigs and chickens he would later slather with barbecue sauce on grandpa's smoker.

After high school, Moss moved from Florence to Columbia with a friend who was attending the University of South Carolina. He started working at Chick-Fil-A, gradually working his way up to management. He knew he wanted to cook but found the doors to the kitchen—any kitchen—firmly shut.

"They always said I didn't have any experience," he said. "Finally a buddy of mine named Jonathan Robinson opened a bar called the Whig, and they needed someone to run their kitchen, so he gave me a shot."

After a few years developing his chops at the Whig, Moss moved to Philadelphia in 2005 with high school buddy Drew Wallace. He stayed there for about a year and a half and then made plans to move to Charleston, South Carolina, with his lady friend, but Robinson and Wallace had other plans for him.

"Drew and Jonathan found this building in Asheville and came up with the Admiral, and they kinda talked me into moving here."

A cinderblock building that looks on both the outside and inside like the dive bar it once was, Moss and his friends soon turned the tiny West Asheville diner into one of the city's best restaurants.

During his stint as executive chef at the Admiral, Moss developed a reputation for intense, flavorful cuisine and creative pairings of fresh ingredients, changing his menus almost daily to take advantage of whatever he found at area tailgate markets and offerings from local farmers who sought out the young chef.

The *New York Times* called the Admiral "one of the hottest tables in town," and soon the Admiral was one of the best-known restaurants in Asheville.

Moss "watched a lot of Food Network as a kid" and came from a family of great cooks who never made it to culinary school. Yet it certainly hasn't hurt him. He's considered one of Asheville's top chefs, as evidenced by his appearance on the semi-finalist list for the 2013 James Beard award.

"I painted a lot as a kid, and food is my new creative outlet," he said. "I've eaten enough food to know what goes together and what doesn't, and cooking is how I express myself."

In 2013, Moss left the Admiral to turn Ben's Tune Up, a former garage on Hilliard Avenue, downtown, into a ramen noodle and sake joint, a venture in which he's partnered with Robinson.

At the same time, he's also opening Buxton Hill, a Southern-style, family-oriented barbecue restaurant, also in partnership with Robinson.

"I grew up eating barbecue," he said. "Going to get a pound of barbecue and white bread was like anyone else going to McDonald's or KFC. It's just what we ate.

"Tourists, chefs and farmers combined make what we have here in Asheville," Moss said. "We don't have the population to support all the restaurants we have without the tourists, and the farmers are awesome about asking us what we need and trying to get it to us."

As to the future of Asheville cuisine, Moss thinks the Southern food trend is only going to intensify. "It's going to be hot. We might be cooking a different version of it ten years from now, but I still think the eye will be on Southern chefs."

Ben's Tune Up is located at 195 Hilliard Avenue, Asheville. Visit www.benstuneup.com.

WILLIAM DISSEN
EXECUTIVE CHEF/OWNER
THE MARKET PLACE

William Dissen had some big shoes to fill when he bought the Market Place, Asheville's iconic fine dining restaurant, from chef/owner Mark Rosenstein in 2009. The West Virginia native not only filled the shoes, he made them his own and took off running.

William Dissen, chef/owner at the Market Place, is a national leader in farm-sourced sustainable cuisine. *Amy Kalyn Sims.*

Dissen grew up feasting on meals his grandmother cooked from the freshly picked vegetables from her farm near Charleston, West Virginia. The connection to food, family and farming is strong in him and drives his cuisine at he Market Place.

"To me, farm to table is a celebration of community; it's a place to get together, to party, to share a meal, to celebrate, to have fun."

His grandmother raised livestock, kept bees, worked a large garden and foraged. His family "basically lived off the land to sustain themselves."

Fruits and vegetables raised in the spring, summer and fall found their way into the root cellar and canning jars to carry them through the winter, Dissen said.

"My grandmother could pull out a jar of green beans in the middle of winter that she had canned the summer before, and they tasted like fresh," Dissen said. "There was bacon fat in everything, and her biscuits were life changing."

Dissen pays homage to the foods his grandmother canned by doing his own for the Market Place, making pickles, pickled vegetables, tomato jam and blackberry preserves during the summer to put a touch of summer on his customers' plates during winter months.

In spite of his food-centric upbringing, being a chef wasn't exactly on his radar. A science buff in high school, his first choice in careers was to go to medical school and become a radiologist. But a high school job as a dishwasher and prep cook at a local country club led to a fascination with the symphony of coordinated chaos that makes up a restaurant kitchen.

At West Virginia University, he earned degrees in French and English while working restaurant jobs and then headed for the prestigious Culinary Institute of America. He also holds a master's degree in international hospitality tourism management from the University of South Carolina.

Dissen cooked in New York City, California and South Carolina before moving to Asheville. "My wife's parents emigrated from India and eventually settled in Waynesville," he said. "When we were looking for a place to open a restaurant, we kept coming back to Asheville."

He linked up with ASAP, met some of the farmers and decided to come here.

"Asheville's a little like everywhere I've lived," Dissen said. "It's a little Appalachian, but when the tourists are here it feels a little like Manhattan and a little like Charleston, South Carolina. It's a unique hodgepodge of people."

It didn't take long after purchasing the Market Place from Rosenstein for Dissen to start garnering a national and local reputation as a creative and innovative chef and a champion of sustainable agriculture and aquaculture. He was named one of "40 Chefs Under 40" by Mother Nature Network, and *Fortune* magazine honored him this year as a "Green Chef" for its annual Fortune Brainstorm Green Conference.

The Market Place is one of only thirty-five restaurants outside Monterey, California, that are direct partners with the Monterey Bay Aquarium's Seafood Watch program. On the local level, he's done benefits for the Appalachian Sustainable Agriculture Project and has been featured on local farm tours.

He also serves on the advisory board of Asheville-Buncombe Technical Community College's Culinary Arts program and works with ASAP's Growing Minds program.

The connections he has made with area farmers are some of the keys to Dissen's success. "Asheville has a unique network of farms and farmers, and we have access to amazing products and that makes our craft easier," he said. "The farmers put so much love and effort into their products that I don't have to manipulate them so much to turn them into something special.

"A lot of our success as chefs in this town is because we have fruits and vegetables that are picked that morning and delivered to our back doors," Dissen said. "You get a level of quality that way that you just can't get any other way."

Dissen sees Asheville's dining scene expanding and growing in the coming years. "We get two to three million tourists a year, and that's predicted to double in the next few years," he said. "I think you'll see people going back to their Southern food roots, but I also think we're going to see everything from dim sum and Vietnamese food to African food."

The Market Place is at 20 Wall Street, Asheville. Call (828) 252-4162 or visit www.marketplacerestaurant.com.

Jason Roy
Executive Chef/Co-Owner
Biscuit Head

Jason Roy didn't waste any time deciding what he wanted to be when he grew up. When most of his classmates at Georgia's Winder-Barrow High School were flipping burgers at McDonald's, Jason was working in the kitchen of a four-star resort, an internship he got through the school's culinary program.

That experience helped him land a culinary scholarship to the Art Institute of Atlanta, where he learned the classic French system from European chefs.

"After graduation, I had two options," Jason said. "I could go to Le Cirque 2000 in New York City, or I had an offer at a resort on the beach in Florida. Being a Georgia boy, I didn't want to go to New York, so I chose the beach."

Despite his aspirations and education, things didn't go as he had planned. "I totally screwed up that job," he remembered. "I didn't know anything about food costs or labor. I just wanted to show off my skills."

Jason Roy came to Asheville after a tenure as executive chef at the famed Greenbriar Hotel. *Amy Kalyn Sims.*

While the pay was great and he "was a nineteen-year-old executive chef making forty-five grand a year," they hadn't told him about the eighty-hour weeks in culinary school. "I got burned out quick," he said. Somehow Roy lucked out, landing a job at the University of Georgia as a traveling chef for an anthropology-geology field program. That took him out West, and he loved it out there.

Then Roy and one of his buddies took off for Colorado, planning to camp and "couch surf for a couple of months." But the day before they were supposed to come back to Georgia, Jason totaled his friend's truck.

"So we just sort of got stuck in Silverthorn, Colorado, and I took a job bartending because I was burned out on cooking. So I became a ski bum for a while," he said.

It didn't take long for the batteries to recharge and the itch to come back. Soon, Jason was back in the kitchen. After a few local kitchen gigs, he got a job at the Alpenglow Stube, a four-star restaurant in nearby Keystone, Colorado. A few jumps later, he was tapped as executive chef at the Greenbriar Restaurant in Boulder.

While he was at Alpenglow Stube, Jason met his wife, Carolyn, who also worked at the same restaurants. Finally, "I wanted to move to Belize and she wanted to move to…well, she didn't really want to move, so we settled on Asheville."

He came into town, answered an ad on Craigslist and ended up as executive chef at Lexington Avenue Brewery.

"LAB was really good to me and for me," Jason said. "It was a good introduction to Asheville and the vibe here. I was even able to set up a program using our grain to feed cows we used at the restaurant, and there's not a lot of places a chef can do those things."

After several years at LAB, running its kitchen and developing a good reputation for his creative cuisine, Jason decided to hang out his shingle and open a breakfast restaurant.

"It just felt like the right thing to do," Roy said. "I thought it was going to take about five years to come together, but it came together in three."

In the spring of 2013, Jason and Carolyn opened Biscuit Head, a Southern-themed, farm-to-fork breakfast restaurant featuring enormous cat head biscuits filled, covered and/or slathered with everything from fried chicken to country ham to tofu and fried green tomatoes. It was an instant hit.

"I contracted with a local farmer for sixty dozen eggs a week, and we went through sixty dozen today," he said. "I'm going to have to re-think the supply chain."

Jason gives Mark Rosenstein much of the credit for Asheville becoming what it has in terms of a food destination. "Mark was the godfather of the farm-to-table movement," he said. "He was responsible for showing everyone that, hey, these farmers are out there and we can work with them and it will be great."

Roy predicts Asheville will head in some diverse directions in the next few years. "I wouldn't count fine dining out, but there's a lot less room for it," he said. "Anything beer-centered is going to do well, as are the ramen place Elliott [Moss] is doing, and barbecue, too. We may even see 'meat and three' country cooking come back. It's going to be interesting."

Biscuit Head is at 733 Haywood Road, Asheville. Call (828) 333-5145 or visit www.biscuitheads.com.

Mike Moore
Executive Chef/Owner
Seven Sows Bourbon & Larder
Blind Pig Supper Club

It would be hard to find a chef anywhere with more of a connection to the land than Mike Moore, the talent and drive behind the Blind Pig Supper Club and Seven Sows Bourbon & Larder. Moore grew up in the sand

Mike Moore, chef/owner of Seven Sows Bourbon & Larder, combines a boyhood spent on the family farm with a top-notch culinary education to turn out stellar Southern cuisine. *Amy Kalyn Sims.*

hills and pine barrens of Eastern North Carolina, on the outskirts of a tiny farming/railroad town called Elm City in Wilson County.

His family has farmed in Wilson County "since before this nation was founded," largely subsistence crops until the movement toward corporate agriculture after World War II made cotton, tobacco and peanuts more profitable.

After his grandfather died unexpectedly, leaving his grandmother with seventy-seven acres of crops to harvest and seven pigs in a pen (hence the name of his restaurant), his grandmother made ends meet working as a butcher, processing hogs at a meatpacking plant.

Although she had to sell a large part of the farm, they still grew their own food. "My mother made lard soap, picked cabbage and turnips from the fields, put up beans and corn and canned tomatoes and made dinner from scratch every night," Moore said.

"This is the world I came from," Mike said. "Food was fresh and healthy, and it was grown and raised by you and it was hard work. Needless to say, I didn't think there was any other way to live; this was all my family had ever known."

After spending seven years as a Raleigh police officer, cooking on his days off to stay sane, Moore decided to turn in his badge and change careers. "I hit the road to San Francisco and started over there. I cut my teeth in the city by the bay and attended culinary school and began traveling as a chef," he said. "I worked at plenty of places up and down Northern California in 2002 and 2003—from Salada Beach Café in Pacifica to Rogue Chefs Culinary Company in Half Moon Bay. I worked several kitchen positions back in my formative years, doing a lot of grunt work," Moore says. "I worked at several restaurants in San Francisco: Aqua, where I worked as garde manger chef and sauté, which earned two Michelin stars in 2004, and Incanto with Chris Cosentino, where I did pretty much any and everything with offal and odd parts of various animals."

Moore's decision to come back to North Carolina was influenced by two factors: "I was broke and very homesick," he said. Living in California made him realize how important his Southern food heritage was to him. "My fondness for the food from where I come from was set on fire," he said. Moore and his wife ended up in Asheville in 2005.

Even though he had returned to his beloved South, all was not rosy for Mike. "I was bored and struggling as a chef. I had closed a restaurant in Greenville, South Carolina, where I was trying to bring the farm-to-table scene into that market," he said. "I was working as an executive chef in a country club serving stale, unadventurous and cliché food from the 1980s to unappreciative, conservative Christian Republicans that were mean as hell."

Thus, the Blind Pig Supper Club was born in response to these desperate times. Blind Pig Supper Club serves monthly pop-up charity dinners with guest chefs and mysterious menus unveiled twenty-four hours in advance. For that matter, the location isn't revealed until a day ahead of time either. And people ate it up. Reservations quickly sold out to the events.

"The Blind Pig was spawned as a Rebel yell to [the country club] experience. It was meant to be just one dinner—that no one knew about," Moore said. "We killed feral pig, nutria and beaver out of the swamps where I grew up and served them in a fine dining and modernized style. It was something that people really enjoyed. They craved adventurous and creative food, and we found our spot."

In March 2013, Moore realized the culmination of all his culinary dreams with the opening of Seven Sows Bourbon & Larder in downtown Asheville.

"Seven Sows is the restaurant concept I've always wanted. It's deeply rooted in North Carolina," Moore said. "The barn wood is from my great-grandfather's old smokehouse, and the tin is off old tobacco barns. It's meant to evoke feeling and place. I feel like I can accomplish a lot of things with the menu in terms of keeping the flame with preserving our cuisine and tapping into many facets of renaissance as well as continuing to push the envelope with modern cuisine and our regional ingredients."

What brought Asheville to its sudden culinary destination status? Moore thinks it's a combination of culinary talent mixed with micro-local food sources.

"We have a thriving tourist industry, and it's getting more and more attention," Moore observed. "There is an amazing proportion of farmers to populace," he said. "Asheville is like Northern California in the '60s and Charleston, South Carolina, in the '90s. It's our time to shine for many, many reasons. People come here for food and culture. Cooks from D.C. to NYC always look at me in amazement and wonder when I tell them I'm from Asheville. That's amazing. We have something very unique here, and we should be proud of that."

Seven Sows Bourbon & Larder is at 77 Biltmore Avenue, Asheville. Call (828) 255-2592 or visit www.sevensows.com. For more on Blind Pig Supper Club, visit blindpigofasheville.com.

REFERENCES CONSULTED

In addition to the sources below, the author relied on in-person, telephone and e-mail interviews with the people profiled in the book as well as his personal archives from articles on restaurants and chefs he wrote for the *Asheville Citizen-Times*, 1998–2008.

Alexander, Bill, et al. "Berkshires, Brahmas and Bees: Agricultural Operations and Farm Life at Biltmore." Unpublished manuscript. Asheville, NC: The Biltmore Company, 2010.

Asheville City Directories. Archived at Special Collections, D.H. Ramsey Library, University of North Carolina at Asheville.

Biltmore Estate and Biltmore Forest. [North Carolina]: Biltmore Estate Co., 1920.

Blake, Barb. "Asheville Chef Dissen Blends Cuisine, Community." *Asheville Citizen-Times*, August 25, 2012.

Buchanan, John O. Oral history, recorded 1994-07-28. Oral History Collection, D.H. Ramsey Library Special Collections, University of North Carolina at Asheville.

Covington, Howard E. *Lady on the Hill: How Biltmore Estate Became an American Icon.* Hoboken, NJ: John Wiley, 2006.

Curatorial Staff, Biltmore Estate. "Biltmore Culinary History: Food and Wine." Unpublished manuscript. Asheville, NC: The Biltmore Company, 2013.

Dykeman, Wilma, and Douglas W. Gorsline. *The French Broad.* New York: Rinehart, 1955.

Glenn, Anne Fitten. *Asheville Beer: An Intoxicating History of Mountain Brewing.* Charleston, SC: The History Press, 2012.

Greenberg, Sue, and Jan Kahn. *Asheville: A Postcard History.* Dover, NH: Arcadia, 1997.

Kephart, Horace. *Our Southern Highlanders.* New York: Outing Pub. Co., 1913.

McQueen, Earlene L. Oral history, recorded 1995-03-16. Oral History Collection, D.H. Ramsey Library Special Collections, University of North Carolina at Asheville.

Strong, Derek Ryan. "An Economic History of Black Business in Asheville, North Carolina, 1921–1951." Unpublished manuscript, 2010. Special Collections, D. H. Ramsey Library, University of North Carolina at Asheville.

Tingle, Billie Walker. Oral history, recorded 2004-02-20. Oral History Collection, D.H. Ramsey Library Special Collections, University of North Carolina at Asheville.

Zourzoukis, Mary. Oral history, recorded 2008-02-19. Oral History Collection, D.H. Ramsey Library Special Collections, University of North Carolina at Asheville.

INDEX

F

G

H

I

J

L

M

O

P

R

S

T

W

Y

ABOUT THE AUTHOR

Rick McDaniel is a food historian, culinary anthropologist and author specializing in the food of the American South. A retired journalist, he covered food and restaurants as a contributing writer for the *Asheville Citizen-Times* from 1998 to 2008.

McDaniel has served as a Southern regional judge for the James Beard Foundation chef and restaurant awards and as a consultant to the producers of *Diners, Drive-Ins and Dives* on Food Network and Anthony Bourdain's *No Reservations* on Travel Channel.

His first book, *An Irresistible History of Southern Food* (The History Press, 2011), is in the reference libraries of Duke University, the University of Chicago and Harvard University.

Visit us at
www.historypress.net

This title is also available as an e-book